Springer Theses

Recognizing Outstanding Ph.D. Research

Aims and Scope

The series "Springer Theses" brings together a selection of the very best Ph.D. theses from around the world and across the physical sciences. Nominated and endorsed by two recognized specialists, each published volume has been selected for its scientific excellence and the high impact of its contents for the pertinent field of research. For greater accessibility to non-specialists, the published versions include an extended introduction, as well as a foreword by the student's supervisor explaining the special relevance of the work for the field. As a whole, the series will provide a valuable resource both for newcomers to the research fields described, and for other scientists seeking detailed background information on special questions. Finally, it provides an accredited documentation of the valuable contributions made by today's younger generation of scientists.

Theses may be nominated for publication in this series by heads of department at internationally leading universities or institutes and should fulfill all of the following criteria

- They must be written in good English.
- The topic should fall within the confines of Chemistry, Physics, Earth Sciences, Engineering and related interdisciplinary fields such as Materials, Nanoscience, Chemical Engineering, Complex Systems and Biophysics.
- The work reported in the thesis must represent a significant scientific advance.
- If the thesis includes previously published material, permission to reproduce this must be gained from the respective copyright holder (a maximum 30% of the thesis should be a verbatim reproduction from the author's previous publications).
- They must have been examined and passed during the 12 months prior to nomination.
- Each thesis should include a foreword by the supervisor outlining the significance of its content.
- The theses should have a clearly defined structure including an introduction accessible to new PhD students and scientists not expert in the relevant field.

Indexed by zbMATH.

More information about this series at http://www.springer.com/series/8790

Olivier Spitz

Mid-infrared Quantum Cascade Lasers for Chaos Secure Communications

Doctoral Thesis accepted by
Institut Polytechnique de Paris, Palaiseau, France

Springer

Author
Dr. Olivier Spitz
Telecom Paris
Institut Polytechnique de Paris
Palaiseau, France

Supervisor
Prof. Frédéric Grillot
Telecom Paris
Institut Polytechnique de Paris
Palaiseau, France

ISSN 2190-5053 ISSN 2190-5061 (electronic)
Springer Theses
ISBN 978-3-030-74306-2 ISBN 978-3-030-74307-9 (eBook)
https://doi.org/10.1007/978-3-030-74307-9

This Springer imprint is published by the registered company Springer Nature Switzerland AG
The registered company address is: Gewerbestrasse 11, 6330 Cham, Switzerland

Supervisor's Foreword

I am very honored that Springer is publishing Olivier Spitz's thesis.

Olivier's Ph.D. investigated non-linear dynamics of quantum cascade lasers (QCLs) operating under external optical feedback (EOF) and the applications thereof such as private communications and neuromorphic computing. QCLs are a new class of unipolar semiconductor lasers based on inter-conduction-band quantum well transitions and that emit light from the mid-infrared (IR) to the THz region. The topic of QCLs is very current and under rapid development. The first QCL was demonstrated in 1994 at Bell Labs, in the mid-IR spectral range, at cryogenic temperatures. Since then, mid-IR-emitting QCLs have matured over the last 15 years hence providing nowadays room-temperature continuous-wave operation across the entire 3-15-μm range, and watts of continuous-wave (CW) output power over the 4–11-μm range. Due to the large number of wavelengths they can achieve, QCLs contribute to many industrial applications such as spectroscopy, optical countermeasures, and free-space communications.

Overall, I have been extremely impressed by the amount of work done by Olivier throughout his Ph.D. and the level of understanding he has achieved. His thesis contains solid theoretic analysis and many pioneer experiments. Many experimental parameters like temperature, external cavity length and bias patterns were investigated, which allowed studying chaos-based phenomena in QCLs under EOF and to precisely control the non-linear dynamics. Also, the influence of the linewidth enhancement factor was carried out in order to understand why QCLs can, experimentally, generate peculiar non-linear dynamics under EOF. His analysis further confirmed that the linewidth enhancement factor (i.e., α-factor) is not as small as previously thought in QCLs. In other words, although the first-order value of the QCL's α-factor is near-zero, any additional perturbation (e.g., thermal effects, spectral hole burning, etc.) produces an increase of the effective α-factor which becomes most likely driven by the second-order terms, a situation well different than that in bipolar lasers.

Olivier's dissertation showed the first proof of concept concerning free-space private communication using a chaotic carrier generated from a mid-IR QCL source. Also, he proved the first chaos synchronization between two mid-IR QCLs across a classical communication channel. The advantages of such method compared to quantum key distribution as well as the potential to improve the performances of the system were described, both in terms of chaos bandwidth and transmission quality. This work constitutes a smart pathway for private optical communication that relies on using the fractal dimension and the complexity of the strange attractor originated from mid-IR QCL sources.

Olivier also investigated the entrainment phenomenon in QCLs under EOF and external periodic bias forcing. The entrainment phenomenon corresponds to the occurrence of a deterministic chaotic pattern composed of frequencies which are linked to the one of the forcing. Results also showed that both the amplitude and the frequency of the forcing play a key role in the number of retrieved spikes per modulation period. In particular, he discovered the birth of extreme events leading to bursts of extra optical power and analyzed them with a thorough statistical analysis. The giant pulses that characterize these extreme events can contribute to sudden, sharp bursts which are key elements for communication in neuromorphic systems. With mid-IR QCLs, Olivier developed a basic optical neuron system operating 10,000 times faster than biological neurons. Fine tuning of modulation and frequency allows control of time intervals between spikes.

Last but not least, the behavior of a QCL subject to cross-polarization feedback and the potential of the emitted all-optical square pattern were discussed. The period and duty cycle of that pattern can be tuned by varying the feedback strength and the polarization of the reinjected beam. He demonstrated an all-optical square modulation, paving the way for integrated photonic memories and direct modulation of light at mid-infrared wavelength.

To summarize, Olivier's thesis provides significant advances toward improving the understanding of QCL dynamics. In particular, initial efforts about free-space transmission were carried out during his Ph.D. thesis. Eventually, the triggering of extreme events in QCLs is of paramount importance for applications such as optical neuron-like systems, which require optical bursts to be triggered in response to a perturbation.

This Ph.D. work unlocks the non-linear dynamics of QCLs and extends the potential applications of this kind of semiconductor lasers in molecular spectroscopy, optical countermeasure systems, and for communication systems inspired by the brain's powerful computational abilities. This excellent work impacts on the large community of chaos-based applications where research is pushed forward to directions with new wavelengths for the realization of private communication systems. Last but not the least, I am sure that this Springer thesis will be of international appeal.

Palaiseau, France
April 2021

Prof. Frédéric Grillot

Abstract

The mid-infrared domain is a promising optical domain because it holds two transparency atmospheric windows, as well as the fingerprint of many chemical compounds. Quantum cascade lasers (QCLs) are one of the available sources in this domain and have already been proven useful for spectroscopic applications and free-space communications. The purpose of that dissertation is to go one step further by implementing a secure free-space communication relying on optical chaos and consequently, to give an accurate cartography of non-linear phenomena in quantum cascade lasers.

Initial efforts about free-space secure chaotic transmission have been carried out during this Ph.D. thesis with two chaos-synchronized QCLs, which is a pioneer result paving the way for mid-infrared private communications. In order to have a global picture about the non-linear dynamics in QCLs under external optical feedback, we tuned many experimental parameters and this allowed us studying new phenomena in QCLs. We thus found similarities between QCLs and laser diodes when the chaotic dropouts are synchronized with an external modulation, known as the entrainment phenomenon. A cross-polarization reinjection technique led to square-wave emission in the output of the QCL. Eventually, we studied the triggering of rogue waves in QCLs. Rogue waves are a quite common phenomenon in optics (among other domains in science) but they have never been triggered on-demand in semiconductor lasers under external optical feedback before. Further studies will try to avoid such phenomenon in the output of a QCL under external optical feedback since it can disturb the message to be transmitted in a secure communication.

All these experimental results allowed a better understanding of the non-linear dynamics of QCLs and will extend the potential applications of this kind of semiconductor lasers, which have currently been restricted to molecular spectroscopy and optical countermeasure systems.

Publications related to this thesis

1. O. Spitz, J. Wu, M. Carras, C.-W. Wong and F. Grillot, "Low-frequency fluctuations of a mid-infrared quantum cascade laser operating at cryogenic temperatures". *Laser Physics Letters*, vol. 15, no. 11, p. 116201, 2018.
2. O. Spitz, J. Wu, M. Carras, C.-W. Wong and F. Grillot, "Chaotic optical power dropouts driven by low frequency bias forcing in a mid-infrared quantum cascade laser". *Scientific Reports*, vol. 9, no. 1, p. 4451, 2019.
3. O. Spitz, J. Wu, A. Herdt, M. Carras, W. Elsäß er, C.-W. Wong and F. Grillot, "Investigation of chaotic and spiking dynamics in mid-infrared quantum cascade lasers operating continuous-waves and under current modulation". *Journal of Selected Toptics in Quantum Electronics*, vol. 25, no. 6, p. 1–11, 2019.
4. O. Spitz, J. Wu, A. Herdt, M. Carras, W. Elsäß er, C.-W. Wong and F. Grillot, "Extreme events in quantum cascade lasers". *Advanced Photonics*, vol. 2, 066001, 2020.

Acknowledgments

Over the past 3 years, I have collaborated with many people in France, United States and Germany, all who have contributed in some way to my experience and ultimately this thesis. If you are not mentioned by name, please see this as a failing on my part and not yours.

Firstly, I would like to thank Prof. Frédéric Grillot. As my principal Ph.D. supervisor at Télécom Paris, his guidance and support have been invaluable throughout these efforts. I thank him for offering me international exchanges in several countries with multiple leading groups in the field, as well as for supporting me on various international conferences which helped me to catch the most recent research topics and technology advances in applied physics. These opportunities were a great experience that will be, without doubt, beneficial to my future career.

I acknowledge the Direction Générale de l'Armement for founding this Ph.D. thesis, and am especially grateful to Dr. Philippe Adam, who helped us link the research with the possible applications.

I am grateful to the research team at mirSense. To Dr. Mathieu Carras, thank you for the opportunity to work in a startup company with cutting-edge technology. I would like to acknowledge the help of Ali Fortas and Dr. Grégory Maisons, for their relevant advice and for the time they spent helping me with the initial experimental setup. I also wish to express my deepest gratitude to Dr. Louise Jumpertz for her valuable work in the field that contributed to the success of my studies.

I would like to thank Prof. Chee-Wei Wong, from University of California Los Angeles, for the opportunity to work with outstanding facilities in an amazing place. Special thanks to Dr. Jiagui Wu for all the issues we solved and all the progress we made about the chaos experiments. The countless hours we spent in the lab were definitely worth it. Thank you for your endless motivation and optimism. All these findings would have not been possible without the material support and advice from Prof. Ben Williams and his students.

I also wish to express my deepest gratitude to Prof. Wolfgang Elsäßer, from Technische Universität Darmstadt, for the fruitful collaboration we carried out. Your expertise in the field was very valuable to me and so was your feedback about my articles and abstracts. Many of them would have not been possible without the priceless help of Andreas Herdt, whose infinite knowledge about Matlab allowed me focusing on the experiments. I wish you all the best for the end of your Ph.D. thesis. If we can be of any assistance, you will always be welcome in our lab.

As I cannot end this acknowledgement without mentioning all my friends, I want to thank my extraordinary group of friends from SupOptique who, through our experiences, made me become the person I am today. Eventually, I would like to thank my family for their support throughout these years.

Contents

Acronyms

AM	Amplitude Modulation
ASE	Amplified Spontaneous Emission
BA	Broad-Area
BER	Bit-Error Rate
CMa	Chaos Masking
CMo	Chaos Modulation
CSK	Chaos Shift Keying
CW	Continuous-Wave
DFB	Distributed Feedback
DIRCM	Direct InfraRed Counter Measures
DOP	Degree Of Polarization
DT	Double Trench
EDFA	Erbium-Doped-Fiber-Amplifier
EDFRL	Erbium-Doped-Fiber-Ring-Laser
EOF	External Optical Feedback
FEC	Forward Error Correction
FM	Frequency Modulation
FP	Fabry-Perot
FSO	Free-Space Optics
FTIR	Fourier Transform InfraRed
FWHM	Full-Width at Half-Maximum
ICL	Interband Cascade Laser
IR	InfraRed
KSE	Kolmogorov-Sinaï Entropy
KYD	Kaplan-Yorke Dimension
LE	Lyapunov Exponent
LEF	Linewidth Enhancement Factor
LFF	Low-Frequency Fluctuations
LIDAR	LIght Detection And Ranging
LIV	Light-Intensity-Voltage
MBE	Molecular Beam Epitaxy
MCT	Mercury-Cadmium-Telluride

NDF	Neutral-Density Filter
NLSE	Non-Linear Schrödinger Equation
NPBS	Non-Polarizing Beam-Splitter
OOK	On-Off Keying
OPO	Optical Parametric Oscillator
PAM	Pulse Amplitude Modulation
PDF	Probability Density Function
QCL	Quantum Cascade Laser
QCW	Quasi-Continuous Wave
QD	Quantum Dot
QEPAS	Quartz Enhanced Photo-Acoustic Spectroscopy
QKD	Quantum Key Distribution
QW	Quantum Well
QWIP	Quantum Well Infrared Photodetector
QWP	Quarter Wave Plate
RADAR	RAdio Detection And Ranging
RAM	Residual Amplitude Modulation
RF	Radio-Frequency
RIN	Relative Intensity Noise
RNG	Random Number Generation
SEM	Scanning Electron Microscopy
SHB	Spatial Hole-Burning
TATP	TriAcetone TriPeroxide
TE	Transverse Electric
TM	Transverse Magnetic
TNT	TriNitroToluene
VCSEL	Vertical-Cavity Surface-Emitting Laser
VdPFN	Van der Pol-Fitzhugh-Nagumo
ZnSe	Zinc-Selenide

Chapter 1
Introduction

This introduction chapter will present the general context of this study, as mid-infrared quantum cascade lasers (QCLs) can be relevant for many applications requiring powerful and versatile mid-infrared optical sources. Extensive details will be given about the QCL opportunities for free-space communications in the following chapters so this introduction will mainly focus on three other aspects: optical countermeasure systems, absorption spectroscopy and lasers for surgery. As QCLs are not the only available sources in the mid-infrared, I will briefly digress on the other options and what their main differences with QCLs are. The organization of the dissertation will furthermore be described.

1.1 The Mid-Infrared Domain

1.1.1 Electromagnetics

An elctromagnetic wave can be described by its wavelength if it is considered as a wave and it can be described by its energy if it is considered as a particle flux [1]. The period of such wave can vary from a few dozens of nanometers for X-rays to several centimeters in the case of radiowaves. The infrared domain ranges from 800 nm, above the visible spectrum, to 20 μm which corresponds to the limit with the terahertz regime. The infrared domain is of prime interest because it corresponds to the wavelength emitted by room-temperature bodies and because it holds two transparency atmospheric windows. Furthermore, mid-infrared wavelength is related to the vibrational absorption of molecules, which means that chemical compounds are able to absorb the energy transmitted by infrared photons.

O. Spitz, *Mid-infrared Quantum Cascade Lasers for Chaos Secure Communications*, Springer Theses, https://doi.org/10.1007/978-3-030-74307-9_1

The blackbody emission, also called Planck's law, provides the relationship between the temperature of that body and the emitted wavelength. It is given by [1]:

$$\rho(\lambda, T) = \frac{2hc^2}{\lambda^5} \frac{1}{e^{\frac{hc}{\lambda k_B T}} - 1} \tag{1.1}$$

where h is the Planck constant, λ the wavelength, c the speed of light, k_B the Boltzmann constant and T the temperature. From this formula, we can derive that an airplane motor will have a maximum emission around 3 μm and a human body will have a maximum emission around 10 μm.

1.1.2 Atmosphere Absorption

The mid-infrared domain holds two transparency windows, one between 3 and 5 μm and another between 8 and14 μm. Consequently, mid-infrared optical sources

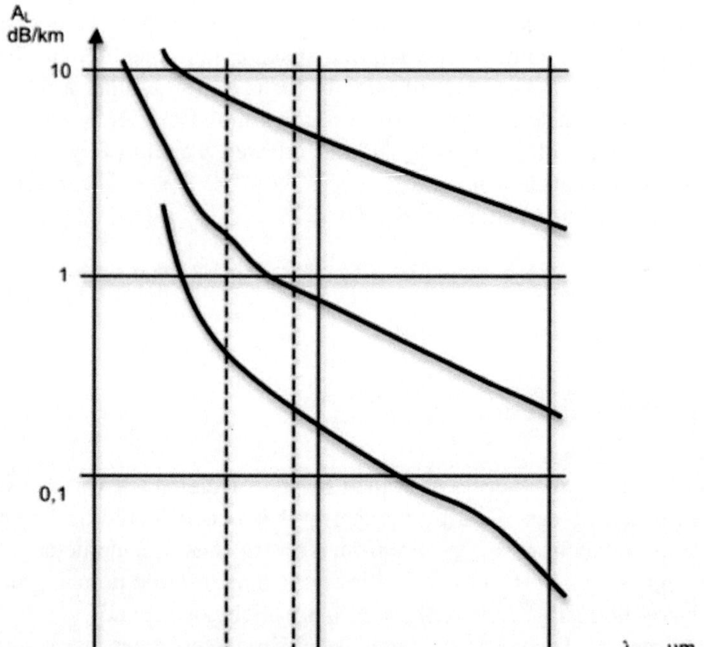

Fig. 1.1 Atmosphere absorption A_L with respect to the wavelength λ. The top line corresponds to fog conditions, the middle line corresponds to regular weather conditions and the bottom line corresponds to high-visibility conditions

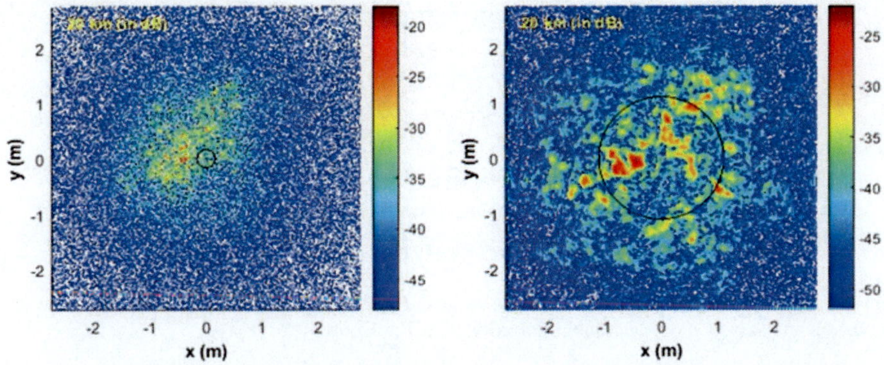

Fig. 1.2 Simulated beam intensity profile in decibel units after a 20 km propagation through atmosphere for 1.55 µm (left chart) and 10.6 µm (right chart) for $C_n^2 = 5 \times 10^{-14}$ m$^{-\frac{2}{3}}$. Solid lines show beam size due to diffraction only. Courtesy of [6]

are candidates of choice for transmission applications, especially in the case of fog or drizzle where they strongly surpass near-infrared and visible-light sources [2], as shown in Fig. 1.1. More details about communications at mid-infrared wavelength will be given in the section dedicated to secure chaotic communications. However, it is relevant to note that not only the absorption but also the signal distortion is lower at higher wavelength. Indeed, it is known that turbulence on the propagation path significantly deteriorates the optical signal causing e.g. beam spreading, beam wandering, scintillation or loss of spatial coherence [3]. In this case, the scintillation will be the predominant phenomenon, corresponding to intensity fluctuations of the propagating beam. This effect evolves as a function of $\lambda^{-7/6}$ [4], and will therefore be less significant at higher wavelength hence the advantage of mid-infrared waves for free-space optical communications. Numerical studies with simulated scintillation data and turbulent strength (also known as C_n^2 [5]) have recently investigated the superiority of mid-infrared wavelength over near-infrared wavelength for long-haul free-space transmissions [6]. Details about the evolution of the beam after a propagation of 20 km for a wavelength of 1.55 µm and for a wavelength of 10.6 µm can be seen in Fig. 1.2.

1.2 Applications at Mid-Infrared Wavelength

1.2.1 Molecular Spectroscopy

The first application in the mid-infrared domain is molecular spectroscopy [7]. Due to their vibrations in the medium, many molecules will absorb light at one specific wavelength, which is considered a fingerprint of the molecule [8]. The presence of

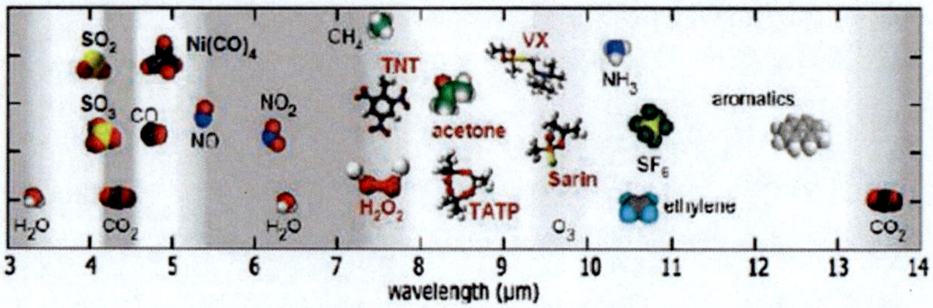

Fig. 1.3 Absorption wavelength of common molecules as well as hazardous molecules such as trinitrotoluene (TNT) and triacetone triperoxide (TATP). Courtesy of [8]

a given molecule can therefore be assessed by illuminating the medium and detecting either directly the optical transmission spectrum, showing absorption peaks at the molecule vibrational frequencies, or through photo-acoustic detection [9, 10], consisting in measuring with microphones the vibration frequency of the molecule, which will be modified due to heating in case of absorption of optical light. The efficiency of this technique can sometimes be improved by the addition of a quartz tuning fork [11, 12], which is known in the literature as quartz enhanced photo-acoustic spectroscopy (QEPAS). The amplitude of the absorption peak or of the acoustic signal is directly related to the amount of the targeted molecule in the studied medium, and molecular spectroscopy can be used to monitor toxic or polluting gases for industrial control. It has also been proven useful in order to detect explosives, drugs or weapons for safety applications [13]. The fundamental vibrational mode of most of these molecules of interest is located in the mid-infrared domain, as shown in Fig. 1.3. It is therefore strategic to illuminate the medium in this range of wavelengths, instead of using the harmonics that can be found in the near-infrared domain, where the spectroscopy would be 2 or 3 orders of magnitude less efficient.

1.2.2 Optical Countermeasure Systems

Optical countermeasure systems [13] is one of the most promising fields of applications for high-power QCLs. They are also known as directed infrared countermeasures (DIRCM) and are designed to protect aircraft from heat-seeking missiles, for example those launched from manportable air defense systems, as illustrated in Fig. 1.4. Depending on the technology, these missiles will operate in the 0.5–2.5 μm range to follow very hot sources such as airplane nozzles, in the 3–5 μm range, in order to follow targets emitting between 300 and 1000 K, or in the 8–14 μm range, to follow cold sources such as naval ships. If they are jammed by a modulated high-power mid-infrared beam, these missiles can be deflected so that they miss their target. This technique is more efficient than the widespread countermeasure flares used by current airplanes. Indeed, the mid-infrared beacon can be used for many

Fig. 1.4 DIRCM system with conventional solid state laser. Courtesy of [13]

times if the aircraft is targeted by several missiles on a raw and the size of packaged QCLs is way smaller than that of a container filled with lures. However, this military application requires state-of-the-art beam quality and robustness which are not accessible to every kind of lasers. For instance, CO and CO_2 lasers were not considered for DIRCM because they do not fully comply with the aforementioned requirements [14]. This is the reason why the first optical sources for countermeasures were optical parametric oscillators (OPO) but the bulkiness of such lasers makes them not as versatile as QCLs [15]. DIRCM require mid-infrared room-temperature lasers with optical power above 1 W and with a beam divergence of less than 5 mrad.

1.2.3 Mid-Infrared Sources for Medical Applications

Many molecules and chemical compounds strongly absorb at mid-infrared wavelength. Thus, mid-infrared lasers are more efficient than their near-infrared counterparts for ablation of biological soft tissues. QCLs emitting around 6 μm have been proven to be of paramount interest for soft tissues coagulation [16], selective removal of atherosclerotic plaque [17], less-invasive laser angioplasty [18] and corneal tissue ablation [19] because they emit in the amide I band and the amide II band of proteins [20]. Moreover, they are more compact than other lasers able to emit at the same wavelength, such as free electron lasers [21] and difference-frequency generation lasers [22]. One of the key parameters that could broaden the use of QCLs for surgery would be an increase of the peak power because medical applications usually require high intensities. Recent achievements with QCLs at 4.8 μm emitting more than 200 W [23] could widespread the technology among the medical domain. Last but not least, QCLs are promising mid-infrared sources for non-invasive determination of glucose, lactate and triglycerides in blood serum [24], which is an improvement compared to conventional methods requiring blood sampling.

1.3 The Mid-Infrared Landscape

1.3.1 Semiconductor Mid-Infrared Sources

Many sources exist in the mid-infrared domain [25], as summarized in Fig. 1.5. Among them, conventional diode lasers are challenged by cascade structures such as quantum cascade lasers (QCLs), for which next chapter is dedicated, and interband cascade lasers (ICLs). Interband cascade lasers are semiconductor lasers exploiting a cascade scheme similar to that of QCLs, but in which the radiative transition occurs between the conduction and valence bands. ICLs are constituted of a succession of interband active areas in order to increase the wall-plug efficiency of the device. The first ICL was proposed in 1995 [26] and then fabricated in 1997 [27]. Since then, the performances of ICLs improved significantly and they have become suitable sources for spectroscopic applications [28]. ICLs can now be operated room-temperature for wavelengths between 2–7 μm and with an output power up to 1 W [29–32]. Implementation of ICLs on silicon [33] could further broaden the applications of such mid-infrared lasers. We will briefly give a definition of type-I and type-II structures as well as their performances.

1.3.2 Type-I Lasers

In type-I quantum well (QW) geometry, electrons and holes are confined within the same material layer [34]. The emission energy is therefore principally determined by the band gap of the quantum well material and the electron and hole confinement

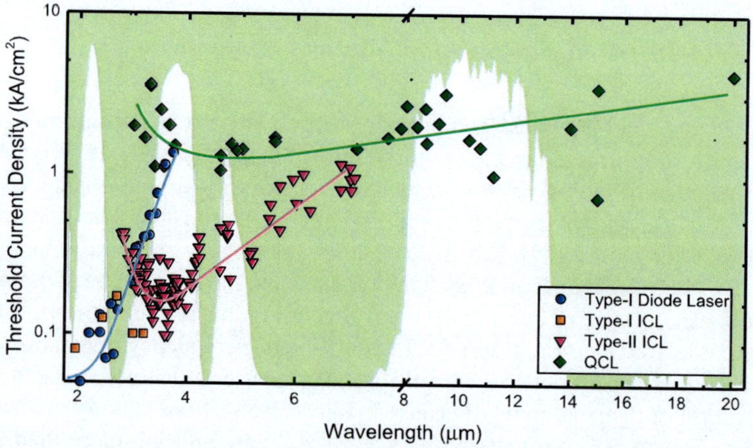

Fig. 1.5 Threshold current density and wavelength of emission for several sources emitting in the mid-infrared domain at room temperature. The white background represents the wavelength ranges where the atmosphere is characterized by a high transparency. The solid lines are for the visual guidance of the reader. Courtesy of [25]

energies. The short wavelength region of the mid-infrared is accessible with conventional type-I quantum well structures. Type-I mid-infrared lasers have achieved room temperature emission in the mid-infrared from 2.0 to 3.73 μm with watt-level output powers out to 2.5 μm [35] and milli-watts at wavelengths beyond 3 μm [36]. The type-I active region has also been incorporated within a cascading scheme [37]. Type-I interband cascade lasers can overcome many of the challenges of conventional lasers offering reduced series resistance, uniform carrier injection and large band offsets. This has enabled lower threshold currents, especially at longer wavelengths, but often at the expense of higher operating voltages [38].

1.3.3 Type-II Lasers

Beyond 3 μm, type-II QW-lasers begin to surpass the performance of the type-I lasers. One approach for such configuration is the type-II superlattice, where the carrier wavefunctions are extended across successive layers. The coupling of the wavefunction between multiple quantum wells forms a miniband and consequently a broad emission spectrum. In the type-II alignment, electrons and holes are confined within separate quantum well layers [39]. This allows for energy transitions less than the bandgap of the constituent materials, extending the accessible wavelengths within a given material system. In addition, the separation of carriers permits independent engineering of the electron and hole quantum wells. On the other hand, the spatial separation of electrons and holes reduces the wavefunction overlap. Type-II ICLs have demonstrated room temperature operation with threshold current densities less than $500\,A.cm^{-2}$ between 2.8 and 5.2 μm [40]. Moving outside this wavelength range, the threshold current density increases rapidly towards both shorter and longer wavelengths. This has been attributed to the efforts that have focused on optimising the active core and waveguide in this region.

1.3.4 CO_2, CO and Lead-Salt Lasers

We will end this section with a short review of the first sources that were considered for mid-infrared applications because they were the first available sources in that domain [41, 42]. Because of their bulkiness and lack of versatility, many of these optical sources are nowadays less commonplace compared to semiconductor mid-infrared lasers, except for very specific applications, like laser cutting of metals [43] or welding of polymers [44]. Among these sources are the CO_2 and CO molecular lasers, which were of prime interest because of their output wavelength in the atmospheric transmission windows near 10 and 5 microns, respectively. CO_2 lasers with efficiencies of more than 30% have been realized [45], while the output power of such lasers can exceed several kW [46]. However, diatomic molecular lasers such as CO offer significant advantages for obtaining operation at even higher efficiencies. Indeed, in the latter case, only one ladder of energy levels is accessible, corresponding to a single vibrational mode (in contrast to CO_2, which has three different modes). The consequence is that it is possible to have a (nearly) 100% quantum efficiency because the lower laser level of a given transition can serve as the upper level of a subsequent transition. This advantage is in contrast to the CO_2 system, for which the energy of the lower laser level is rapidly dissipated as heat energy through vibration-translation collisional relaxations and that prevents high quantum efficiencies. Optical power up to 70 W was achieved with a CO laser at cryogenic temperature [47], with the view of using it for spectroscopic applications [48].

The last mid-infrared laser sources to be introduced are lead-salt lasers. These semiconductor lasers based on PbSe, PbTe or PnS have a direct bandgap, but the transition does not occur at the Γ point (i.e. at the center of the Brillouin zone), but at one of the points corresponding to off-centered band minima. Moreover, the effective mass of the electrons and holes are similar, and consequently, there are only few Auger recombinations, leading to a possible population inversion and laser effect. The achievable wavelength ranges from 3 to 30 μm depending on the addition of Cd, Sn, Eu or Yb [49, 50]. These lasers emit typically a few hundreds of μW, but they are limited in their applications because of their restricted reliability and need for cryogenic operation. Furthermore, the performances of such devices are strongly unstable with a small temperature fluctuation.

1.4 Organization of the Dissertation

The principle of operation of QCLs will be explained in the next chapter. A theoretical analysis will provide details about the specificity of this kind of semiconductor lasers. This will emphasize the differences existing between QCLs and the widespread near-infrared laser diodes. In particular, the influence of the linewidth enhancement factor will be detailed in order to understand why QCLs can, experimentally, generate non-linear dynamics under external optical feedback.

External optical feedback will be thoroughly described in the third chapter, after a general introduction about chaos, the phase portraits and bifurcation diagrams. External optical feedback consists in reinjecting part of the emitted light back into the laser. Optical feedback has been widely studied in interband laser diodes, where it can either improve or deteriorate the laser properties depending on the two main parameters that are the external-cavity length, or feedback delay, and the feedback strength, defined as the ratio between reinjected and emitted powers. Examples of the external optical feedback's positive and negative effects will be given in the specific case of the newly investigated QCLs non-linear dynamics. The dissertation will then focus on several experimental parameters such as temperature, external-cavity length and bias patterns in order to precisely control the chaotic dynamics found in the output of the lasers. The entrainment phenomenon will be introduced and the influence of the amplitude and frequency of the modulation will be detailed. The behavior of a QCL subject to cross-polarization feedback will be discussed and the potential of the emitted all-optical square-wave pattern will be explained.

The fourth chapter will focus on the chaos synchronization between two semiconductor lasers. A simple theoretical example will be given, associated with a review of the first successful chaos synchronizations in various domains. The usefulness of synchronization will then be characterized through the prism of secure communication that was achieved for the first time more than 20 years ago. Yet, this type of transmissions remains limited to optical fiber systems. We will show initial experiments concerning free-space secure communication in the mid-infrared and will give the advantages of such method compared to quantum key distribution. Eventually, we will describe what are the potential levers in order to improve the performances of our current system, both in terms of chaos bandwidth and transmission quality. The concept of optical injection will be introduced and the non-linear dynamics obtained with that scheme will be compared with those retrieved with conventional optical feedback. Opportunities about mixing optical injection and conventional optical feedback will also be presented.

As a new non-linear phenomenon in QCLs with external optical feedback, rogue waves and extreme events will be the cornerstone of chapter five. The conditions that favor such events will be described and a thorough statistical analysis will be performed in order to confirm the existence of rogue waves. As extreme events represent bursts of extra optical power, a method will be given in order to trigger them precisely and that is compatible with sensing applications. Tools to predict such events will also be introduced with the view of hindering rogue waves in the cases where they could be hazardous. Eventually, peculiarities of extreme events such as dragon kings will also be briefly discussed and comparison with other fields of physics will be drawn.

Finally, the last chapter will highlight the main discoveries of this work and discuss some perspectives for future experimental and numerical studies.

References

1. Feynman R, Leighton R, Sands M (2013) Le cours de physique de Feynman. Dunod
2. Colvero C, Cordeiro M, Von der Weid J (2005) Real-time measurements of visibility and transmission in far-, mid-and near-ir free space optical links. Electron Lett 41(10):610–611
3. Delga A, Leviandier L (2019) Free-space optical communications with quantum cascade lasers. In: Quantum sensing and nano electronics and photonics XVI, vol 10926. International Society for Optics and Photonics, p 1092617
4. Majumdar AK, Ricklin JC (2010) Free-space laser communications: principles and advances, vol 2. Springer Science & Business Media
5. Otoniel Canuet LF (2015) Atmospheric turbulence profile modeling for satellite-ground laser communication, Master's thesis, Universitat Politècnica de Catalunya
6. Liu JJ, Stann BL, Klett KK, Cho PS, Pellegrino PM (2019) Mid and long-wave infrared free-space optical communication. In: Laser communication and propagation through the atmosphere and oceans VIII, vol 11133. International Society for Optics and Photonics, p 1113302
7. Lee BG, Belkin MA, Audet R, MacArthur J, Diehl L, Pflügl C, Capasso F, Oakley DC, Chapman D, Napoleone A et al (2007) Widely tunable single-mode quantum cascade laser source for mid-infrared spectroscopy. Appl Phys Lett 91(23):231101

8. Takeuchi EB, Rayner T, Weida M, Crivello S, Day T (2007) Standoff detection of explosives and chemical agents using broadly tuned external-cavity quantum cascade lasers (EC-QCLs). In: Optics and photonics for counterterrorism and crime fighting III, vol 6741. International Society for Optics and Photonics, p 674104
9. Kosterev A, Wysocki G, Bakhirkin Y, So S, Lewicki R, Fraser M, Tittel F, Curl R (2008) Application of quantum cascade lasers to trace gas analysis. Appl Phys B 90(2):165–176
10. Coutard JG, Glière A, Fedeli J-M, Lartigue O, Skubich J, Aoust G, Teulle A, Strahl T, Nicoletti S, Carras M et al (2019) Photoacoustic cell on silicon for mid-infrared QCL-based spectroscopic analysis. In: MOEMS and miniaturized systems XVIII, vol 10931. International Society for Optics and Photonics, p 109310V
11. Kosterev AA, Bakhirkin YA, Curl RF, Tittel FK (2002) Quartz-enhanced photoacoustic spectroscopy. Opt Lett 27(21):1902–1904
12. Duquesnoy M, Aoust G, Melkonian J-M, Lévy R, Raybaut M, Godard A (2019) Quartz enhanced photoacoustic spectroscopy based on a custom quartz tuning fork. Sensors 19(6):1362
13. Grasso RJ (2016) Defence and security applications of quantum cascade lasers. In: Optical sensing, imaging, and photon counting: nanostructured devices and applications 2016, vol 9933. International Society for Optics and Photonics, p 99330F
14. Abramov P, Kuznetsov E, Skvortsov L (2017) Prospects of using quantum-cascade lasers in optoelectronic countermeasure systems. J Opt Technol 84(5):331–341
15. Bekman HT, Van Den Heuvel J, Van Putten F, Schleijpen R (2004) Development of a mid-infrared laser for study of infrared countermeasures techniques. In: Technologies for optical countermeasures, vol 5615. International Society for Optics and Photonics, pp 27–38
16. Hashimura K, Ishii K, Akikusa N, Edamura T, Yoshida H, Awazu K (2014)Coagulation and ablation of biological soft tissue by quantum cascade laser with peak wavelength of 5.7 μ m. J Innov Opt Health Sci 7(03):1450029
17. Hashimura K, Ishii K, Awazu K (2015) Selective removal of atherosclerotic plaque with a quantum cascade laser in the 5.7 μm wavelength range. Jpn J Appl Phys 54(11):112701
18. Hashimura K, Ishii K, Akikusa N, Edamura T, Yoshida H, Awazu K (2012) Irradiation effects on cholesteryl ester and porcine thoracic aorta of quantum cascade laser in 5.7-μm wavelength range for less-invasive laser angioplasty. Adv Biomed Eng 1:74–80
19. Huang Y, Kang JU (2012) Corneal tissue ablation using 6.1 μm quantum cascade laser. In: Ophthalmic technologies XXII, vol 8209. International Society for Optics and Photonics, p 82091W
20. Edwards G, Logan R, Copeland M, Reinisch L, Davidson J, Johnson B, Maciunas R, Mendenhall M, Ossoff R, Tribble J et al (1994) Tissue ablation by a free-electron laser tuned to the amide II band. Nature 371(6496):416–419
21. Fukami Y, Awazu K (2003) The thermal dissociation of cholesterol esters using a 5.75-μm-free electron laser. Jpn J Appl Phys 42(6R):3716
22. Ishii K, Tsukimoto H, Hazama H, Awazu K (2008) Selective removal of cholesterol ester in atherosclerotic plaque using nanosecond pulsed laser at 5.75 μm. In: Optical interactions with tissue and cells XIX, vol 6854. International Society for Optics and Photonics, p 685418
23. Razeghi M (2020) High power, high wall-plug efficiency, high reliability, continuous-wave operation quantum cascade lasers at Center for Quantum Devices. In: Optical, opto-atomic, and entanglement-enhanced precision metrology II, vol 11296. International Society for Optics and Photonics, p 112961C
24. Brandstetter M, Volgger L, Genner A, Jungbauer C, Lendl B (2013) Direct determination of glucose, lactate and triglycerides in blood serum by a tunable quantum cascade laser-based mid-IR sensor. Appl Phys B 110(2):233–239
25. Eales T (2019) Influence of electronic band structure on the efficiency of mid-infrared photonic devices. PhD dissertation, University of Surrey
26. Yang RQ (1995) Infrared laser based on intersubband transitions in quantum wells. Superlattices Microstruct 17(1):77–83
27. Lin C-H, Yang RQ, Zhang D, Murry S, Pei S, Allerman A, Kurtz S (1997) Type-II interband quantum cascade laser at 3.8 μm. Electron Lett 33(7):598–599

28. Li C, Dong L, Zheng C, Tittel FK (2016) Compact TDLAS based optical sensor for ppb-level ethane detection by use of a 3.34 μm room-temperature CW interband cascade laser. Sens Actuators B Chem 232:188–194
29. Shterengas L, Liang R, Kipshidze G, Hosoda T, Belenky G, Bowman SS, Tober RL (2014) Cascade type-I quantum well diode lasers emitting 960 mW near 3 μm. Appl Phys Lett 105(16):161112
30. Dallner M, Hau F, Höfling S, Kamp M (2015) InAs-based interband-cascade-lasers emitting around 7 μm with threshold current densities below 1 kA/cm^2 at room temperature. Appl Phys Lett 106(4):041108
31. Höfling S, Weih R, Dallner M, Scheuermann J, von Edlinger M, Nähle L, Fischer M, Koeth J, Kamp M (2015) Mid-infrared ($\approx$ 2.8μm to $\approx$ 7.1μm) interband cascade lasers. In: Biosensing and nanomedicine VIII, vol 9550. International Society for Optics and Photonics, p 95500F
32. Hosoda T, Feng T, Shterengas L, Kipshidze G, Belenky G (2016) High power cascade diode lasers emitting near 2 μm. Appl Phys Lett 108(13):131109
33. Spott A, Stanton EJ, Torres A, Davenport ML, Canedy CL, Vurgaftman I, Kim M, Kim CS, Merritt CD, Bewley WW et al (2018) Interband cascade laser on silicon. Optica 5(8):996–1005
34. Krier A (2007) Mid-infrared semiconductor optoelectronics, vol 118. Springer
35. Kim J, Shterengas L, Martinelli R, Belenky G, Garbuzov D, Chan W (2002) Room-temperature 2.5 μm InGaAsSb/AlGaAsSb diode lasers emitting 1 W continuous waves. Appl Phys Lett 81(17):3146–3148
36. Shterengas L, Belenky G, Kipshidze G, Hosoda T (2008) Room temperature operated 3.1 μm type-I GaSb-based diode lasers with 80 mW continuous-wave output power. Appl Phys Lett 92(17):171111
37. Jiang Y, Li L, Yang RQ, Gupta JA, Aers GC, Dupont E, Baribeau J-M, Wu X, Johnson MB (2015) Type-I interband cascade lasers near 3.2 μm. Appl Phys Lett 106(4), 041117
38. Shterengas L, Kipshidze G, Hosoda T, Wang M, Feng T, Belenky G (2016) Cascade type-I quantum well GaSb-based diode lasers. In: Photonics, vol 3, no 2. Multidisciplinary Digital Publishing Institute, p 27
39. Kroemer H, Griffiths G (1983) Staggered-lineup heterojunctions as sources of tunable below-gap radiation: operating principle and semiconductor selection. IEEE Electron Device Lett 4(1):20–22
40. Vurgaftman I, Weih R, Kamp M, Meyer J, Canedy C, Kim C, Kim M, Bewley W, Merritt C, Abell J et al (2015) Interband cascade lasers. J Phys D Appl Phys 48(12):123001
41. Sobolev N, Sokovikov VV (1967) CO_2 lasers. Sov Phys Uspekhi 10(2):153
42. Nussmeier T, Goodwin F, Zavin J (1974) A 10.6 μm terrestrial communication link. IEEE J Quantum Electron 10(2):230–235
43. Stournaras A, Stavropoulos P, Salonitis K, Chryssolouris G (2009) An investigation of quality in CO_2 laser cutting of aluminum. CIRP J Manuf Sci Technol 2(1):61–69
44. Duley W, Mueller R (1992) CO_2 laser welding of polymers. Polym Eng Sci 32(9):582–585
45. Dezenberg G, Merritt J (1967) The use of a multipath cell as a CO_2-N_2 gas laser amplifier and oscillator. IEEE J Quantum Electron 3(6):268–269
46. Afonin YV, Golyshev AP, Ivanchenko A, Malov AN, Orishich AM, Pechurin VA, Filev VF, Shulyat'ev VB (2004) High-quality beam generation in a 8-kW CW CO_2 laser. Quantum Electron 34(4):307
47. Bhaumik M, Lacina W, Mann M (1972) Characteristics of a CO laser. IEEE J Quantum Electron 8(2):150–160
48. Urban W (1991) The carbon monoxide laser as spectroscopic source. Laser Optoelektron 23:56–61
49. Partin DL (1988) Lead salt quantum effect structures. IEEE J Quantum Electron 24(8):1716–1726
50. Tacke M (1995) New developments and applications of tunable IR lead salt lasers. Infrared Phys Technol 36(1):447–463

Chapter 2
Quantum Cascade Lasers: Mid-Infrared Sources with Outstanding Features

This chapter aims at describing the properties and conditions of operation of quantum cascade lasers. A short review of the technology progress will be presented with examples for both mid-infrared QCLs and Terahertz QCLs. A specific focus will be dedicated to the linewidth enhancement factor, that is a key parameter for the analysis of non-linear dynamics in semiconductor lasers. Two methods will be underlined to retrieve the linewidth enhancement factor above threshold and below threshold, respectively. This chapter will end with two examples of external optical feedback application other than laser output destabilization. These two cases concern beam steering of multi transverse-modes QCLs and the Talbot coupling QCL arrays.

2.1 From Theory to The Device

2.1.1 Mid-Infrared Quantum Cascade Lasers

QCLs are optical sources exploiting radiative intersubband transitions within the conduction band of semiconductor heterostructures which are called superlattices (Fig. 2.1). A superlattice is an artificial semiconductor material consisting of a large number of periods, and each period holds two layers of dissimilar materials presenting different gap energies but similar lattice constants. Roughly ten years after the laser discovery, Kazarinov and Suris were the first ones to introduce a population inversion based on semiconductor intersubband transitions in order to generate a stimulated emission [1]. Following this discovery, tunnelling in superlattices [2], resonant tunnelling in semiconductor double barriers [3] and other transport properties of superlattices [4] were extensively studied, leading to the experimental proof

O. Spitz, *Mid-infrared Quantum Cascade Lasers for Chaos Secure Communications*, Springer Theses, https://doi.org/10.1007/978-3-030-74307-9_2

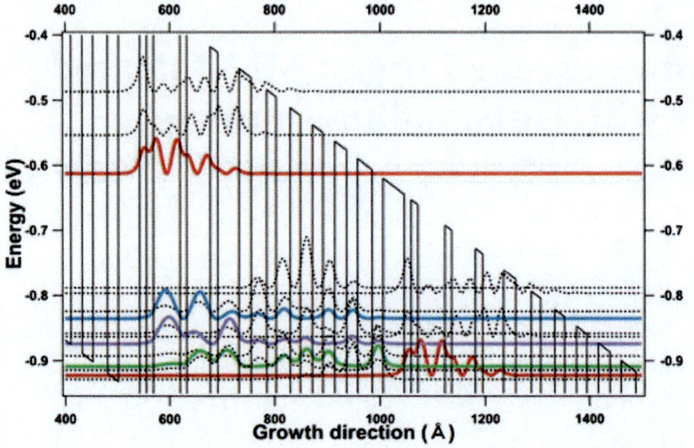

Fig. 2.1 Example of superlattice modelling with several heterostructure widths leading to optical emission in the mid-infrared. Wave-functions of interest are stressed with colors and the emission process will be explained in the following

of discretization in quantum wells [5], though QCLs had not been produced yet. One of the cornerstones of the experimental fabrication of QCLs was the development of growing techniques such as molecular beam epitaxy (MBE) [6] and band-gap engineering [7]. This led to the first QCL in 1994 [8], within the Bell Labs facilities after nearly one decade of unfruitful trials. As we already saw that QCLs are an optical source of choice in the mid-infrared domain, they were quickly envisioned for many applications hence drawing attention on practical improvements. The first QCL was indeed pumped at a maximum temperature of 88 K and could only be operated in pulsed mode. A few months after the first experimental demonstration, new discoveries brought QCLs operated in CW mode at cryogenic temperatures [9] and in pulsed mode up to room temperature [10], as well as new geometries. One can cite, for instance, distributed feedback (DFB) QCLs [11], providing continuously tunable single-mode laser output, quantum cascade disk lasers [12] and quantum cascade "bow-tie" lasers [13]. Simultaneously, the quest for higher power and longer-wavelength devices led to the invention of the superlattice active region [14]. Rather than providing population inversion between subbands through appropriate design, minibands in active region superlattices allow achieving an intrinsic inversion and this authorizes large injection currents. The latter is intrinsically related to large optical power, which is typically obtained for large pump currents, and long wavelengths, where the larger losses require an active region able to sustain a large threshold current. Further improvements have then been performed by several groups [15–18], leading to powerful and versatile optical sources between 2.75 and 10.6 μm. Examples of outstanding QCLs working in CW mode are displayed in Table 2.1, for a selected number of wavelengths. Recent insights allowed envisioning further improvements of the QCL technology and its potential for large-scale adoption with the development of QCLs on silicon [19, 20] and quantum dot cascade lasers [21]

Table 2.1 Experimental parameters of outstanding QCLs as estimated or measured and published in the literature

References	[23]	[24]	[25]	[26]	[27]	[28]
Wavelength (μm)	3.4	4	4.6	7.1	9	10.6
Output power (W)	0.4	2	5.1	1.2	2	0.1

with the purpose of taking advantage of slowing of the relaxation processes between confined state in quantum dot (QD) lasers [22].

2.1.2 Terahertz Quantum Cascade Lasers

The emitting wavelength of QCLs has been pushed up to 24 μm [29] in the case of surface-plasmon interactions, and this corresponds to the upper limit for mid-infrared QCLs. Indeed, the wavelength range from 28 to 45 μm corresponds to the Reststrahlen band, where the III–V materials are absorbant due to optical phonon absorption and consequently, no QCLs based on these materials can emit in this domain. However, it is possible to produce QCLs emitting above 45 μm or, equivalently, below 6 THz. These lasers are called terahertz (THz) QCLs [30] and until now, the maximum temperature for CW operation is 129 K [31], because of the appearance of thermal relaxation mechanisms between upper and lower laser levels through optical phonons, preventing from population inversion when increasing the temperature. In order to increase the temperature of operation of such QCLs, ingenious schemes have been developed, such as the application of a strong magnetic field above 16 T, in order to suppress the inter-Landau-level non-radiative scattering [32]. Another scheme based on difference frequency generation in a mid-IR QCL to obtain monolithic THz devices has been proposed [33] and has led to THz room-temperature sources with high output power. QCLs emitting in the THz domain are of prime interest for spectroscopy of explosives and drugs [34] as well as for real-time imaging [35]. Figure 2.2 presents the QCL performances reported in the literature [36], over the mid-infrared and THz ranges from 2.75 to 320 μm for three conditions of operation: CW operation, pulsed operation and magnetic-field assisted operation.

2.1.3 Charge Transport in QCLs

The emitting process of QCLs relies on electron relaxation from a conduction subband toward another subband from the same conduction band. The mid-infrared wavelength is thus not fixed by the semiconductor bandgap, contrary to what happens with interband lasers. Intersubband transitions require a complex engineering

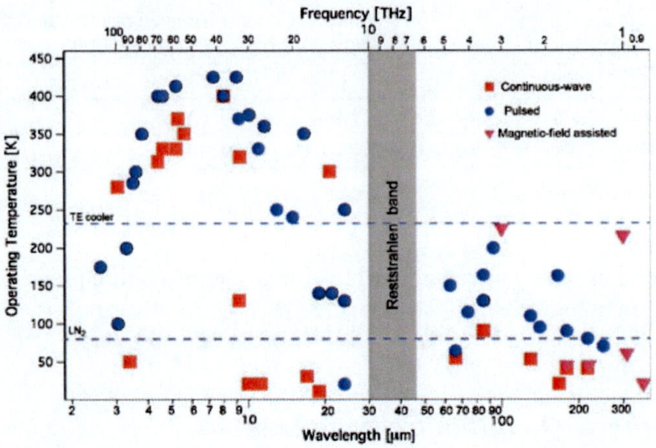

Fig. 2.2 Operating temperature plot as a function of the emission wavelength/frequency for mid-infrared and terahertz quantum cascade lasers. Lasers are gathered according to their principle of operation (continuous-wave, pulsed or magnetic-field assisted). Courtesy of [36]

of the energy levels in the quantum wells of the superlattice. The only limitation in wavelength is the thickness of the quantum well where the laser transition takes place, which rules the spacing between the subbands. On the one hand, if the upper subband is too close to the continuum, the electron will no longer be confined and no photon will be emitted if the quantum well is too narrow. On the other hand, the subbands will be very close from one another and thermal relaxation will compete with the radiative transitions if the quantum well is too wide. This is the reason why the aforementioned THz QCLs are limited to temperatures below 225 K, in pulsed mode. Figure 2.3 shows the typical structure of a QCL as well as details of the active area. The latter is composed of several periods and each period is made of an alternation of wells (white lines) and barriers (black lines). This combination of wells and barriers gives rise to an emission area where stimulated emission occurs [37], followed by an injection area which links period N-1 with period N by emptying the injection level of period N-1, as summarized in Fig. 2.4. A basic way to model the behavior of one period is to consider a simple-level analysis of the emission area. This area is assumed to consist of three states with level $|3\rangle$ the upper level of the emission area, level $|2\rangle$ the lower one and level $|1\rangle$ the injection level from period N-1 towards level $|3\rangle$ from period N. Level $|1\rangle$ is assumed to have a constant population n_g, aligned with the upper level and the next period. Electrons are injected at level $|3\rangle$ at a rate equal to $\frac{J}{q}$. Electrons may scatter from this level to the aforementioned lower levels. The lifetime of level $|3\rangle$, called τ_c or carrier lifetime, can be approximated by [38]:

$$\frac{1}{\tau_c} \approx \frac{1}{\tau_{31}} + \frac{1}{\tau_{32}} \tag{2.1}$$

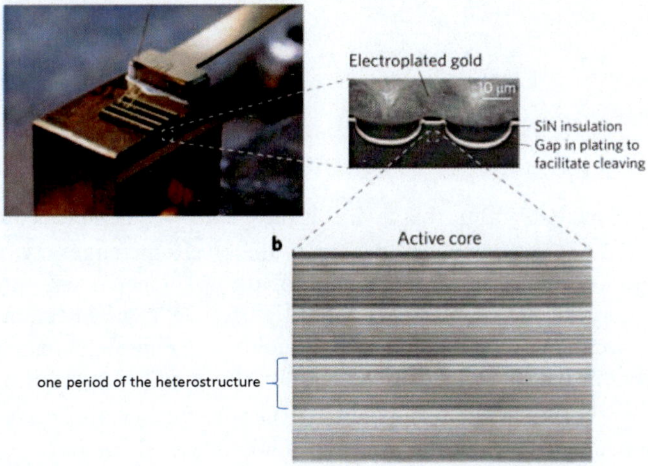

Fig. 2.3 Photograph of a laser bar with four QCL ridges (top left) and scanning electron microscopy (SEM) image of the front facet of one of the QCLs (top right). High-resolution close-up of the front facet achieved with a transmission electron microscope (bottom right). One period of the heterostructure is emphasized with the alternation of wells (white lines) and barriers (black lines). Courtesy of [37]

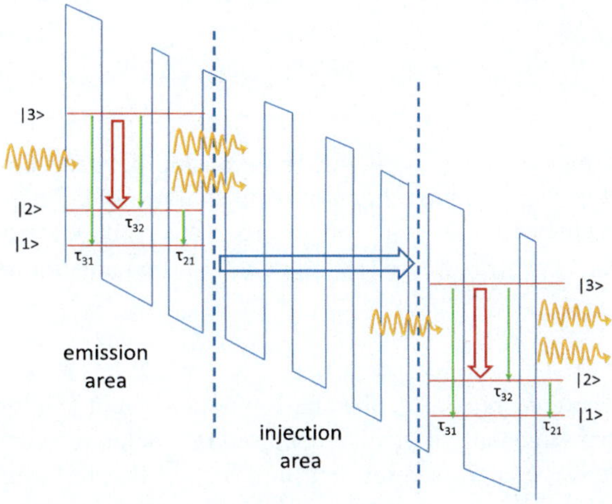

Fig. 2.4 Simplified schematic of the conduction band structure for a QCL, where the laser transition is between sub-bands 3 and 2

if we get rid of the carrier escape into the continuum. The set of rate equations describing the change of the populations is [39]:

$$\frac{dn_3}{dt} = \eta\frac{J}{q} - \frac{n_3}{\tau_c} - Sg(n_3 - n_2) \tag{2.2}$$

$$\frac{dn_2}{dt} = (1-\eta)\frac{J}{q}\frac{n_3}{\tau_{32}} + Sg(n_3 - n_2) - \frac{n_2 - n_{th}}{\tau_2} \tag{2.3}$$

$$\frac{dS}{dt} = \frac{c}{n_{eff}}((Sg(n_3 - n_2) - S\alpha_{tot}) + \frac{\beta n_3}{\tau_{sp}}) \tag{2.4}$$

with S the photon flux within the cavity, g the gain, $\frac{c}{n_{eff}}$ the speed of photon in the material, J the injection current density, q the electron charge, α_{tot} the sum of waveguide and mirror losses, β the fraction of the spontaneous emission coupled in the laser mode, τ_{sp} the spontaneous emission lifetime, η the injection efficiency from $|1\rangle$ in a given emission area to $|3\rangle$ in the following emission area (symbolized by a blue arrow in Fig. 2.4) and n_{th} the equilibrium population at level $|2\rangle$ which scales $n_{th} = n_g e^{\frac{-\Delta}{k_B T}}$ with n_g the sheet doping density of the injector and Δ the energy difference between the Fermi level of the injector and level $|2\rangle$ of the laser transition.

2.1.4 Laser Properties

From the previous three equations, it is possible to derive the threshold current of the QCL, which is the bias current required so that the gain compensate the losses:

$$J_{th} = \frac{q(\frac{\alpha_{tot}}{g} + n_{th})}{\eta\tau_{eff} - \tau_2(1-\eta)} \tag{2.5}$$

with $\tau_{eff} = \tau_c(1 - \frac{\tau_2}{\tau_{32}})$ the relationship between the population inversion and the electrical pumping. In order to reach threshold, the injection current must overcome the losses of the cavity and the losses due to thermal backfilling. The threshold current can also be expressed as a function of the temperature of the active area T_a:

$$J_{th} = J_0 e^{\frac{T_a}{T_0}} \tag{2.6}$$

where J_0 and T_0 are the characteristic current and temperature of the QCL and are experimentally derived in a pulsed bias scheme in order to avoid overheating of the active area. T_0 is a relevant indicator of the performances of the QCL under study and evaluates the stability of the threshold current when varying the temperature. Values between 150 and 300 K are commonplace for T_0 [40], and $T_0 = \infty$ would mean that the laser is totally insensitive to temperature changes. It is also possible to write the photon flux as a function of the injection current:

$$S = \frac{(J - J_{th})\big(\tau_{eff}\eta - \tau_2(1-\eta)\big)}{q\alpha_{tot}(\tau_2 + \tau_{eff})} \tag{2.7}$$

which leads to the slope efficiency for the whole stack:

$$\frac{dP}{dI} = \frac{1}{2}\frac{\hbar\omega}{q}N_p\frac{\alpha_M}{\alpha_{tot}}\frac{\tau_{eff}\eta - \tau_2(1-\eta)}{\tau_2 + \tau eff} = \frac{1}{2}\frac{\hbar\omega}{q}N_p\frac{\alpha_M}{\alpha_{tot}}\eta_{int} \tag{2.8}$$

with α_M the front mirror loss, N_p the number of periods and η_{int} the internal quantum efficiency. α_M can be written:

$$\alpha_M = \frac{ln(R_1 R_2)}{2L} \tag{2.9}$$

where R_1 and R_2 are the facet reflectivities and L is the length of the waveguide. Throughout the process, losses can be explained by either the interaction between electrons and photons, i.e. free-carriers absorption, or by scattering because of materials defects or waveguide sidewall roughness [41].

Two main kinds of structure exist in QCL manufacturing. The first realisations relied on double-channel ridge waveguides in which the active area is split in three parts, separated by isolating trenches. A metal deposition on top allows biasing the structure [42]. The main advantage of this process is that leak currents are almost non-existent but the efficiency of the heat sink is not optimized and it is tricky to achieve CW operation in that case. The other option is called buried heterostructure, as illustrated in Fig. 2.5. The active area is etched on an InP substrate and then, an Iron doped Indium Phosphide regrowth is realized in order to maximize thermal dissipation [43]. In the following, details will be given about the QCLs which were used in our EOF experiments. All the QCLs we used had a DFB structure and came from the same batch, which means that small changes can be found between two lasers but their overall properties remain the same. This is the reason why only some of the lasers can be operated under CW bias at room temperature. A DFB structure

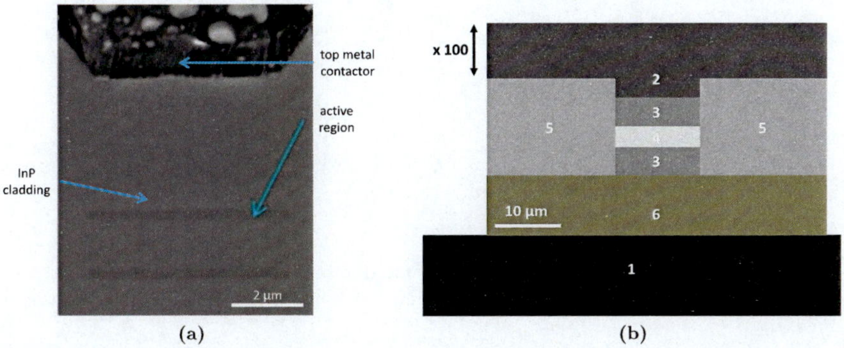

Fig. 2.5 **a** Scanning electron microscope picture of a buried heterostructure QCL. **b** Schematic of the emission facet of a QCL epi-side down mounted with mounting base (1), substrate (2), cladding (3), active region (4), dielectric region (5) and gold metalization (6). The substrate is one hundred times wider than shown on the schematic and the mounting base can be a few millimeters wide and several centimeters large

allows emitting single-mode, contrary to FP-lasers which have a multi-mode optical spectrum.

The DFB-QCLs were manufactured by mirSense. They emit single-mode radiation around 5.6 µm with 30 dB side-mode suppression and have a wavelength shift of 20 nm.A^{-1}, in average. The lasers are tunable mode-hop free by actuating on the operation temperature and current, with a maximum output power up to 180 mW at room temperature. Figures 2.6 and 2.7 show the optical spectrum and the LIV curves of one of the QCLs of interest. The devices under study were grown by molecular beam epitaxy on an InP cladding and incorporate 30 periods of AlInAs/GaInAs layers [16]. The upper InP cladding is n-doped at a value of 10^{17} cm^{-3} in order to get electrical injection but without introducing any plasmonic effects. The design for the DFB-lasers uses index coupling and metal grating, as reported in Refs. [44, 45]. This enables single-mode emission using a top metal grating with a coupling efficiency of $\kappa \approx 4\,\text{cm}^{-1}$. The QCLs emit a continuous wave thanks to a standard double-trench (DT) configuration without Iron doped Indium Phosphide regrowth. The combination of the top metal grating approach and the DT configuration ensures a stable and reproducible process. The back-facet of the QCLs has a high-reflectivity coating (>95%) while the front facet is left as-cleaved.

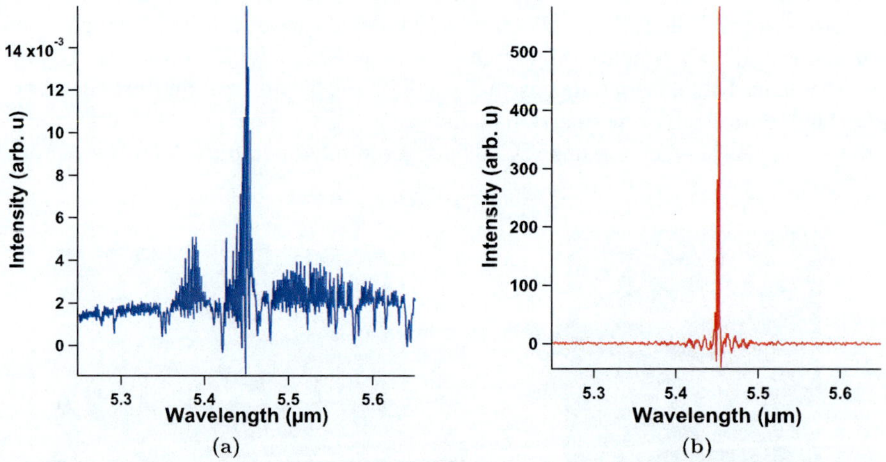

Fig. 2.6 Typical spectral characteristics of one of the free-running DFB-QCLs under study. The laser is operated at 77 K and under a continuous bias 5 mA below threshold (**a**) and 5 mA above threshold (**b**)

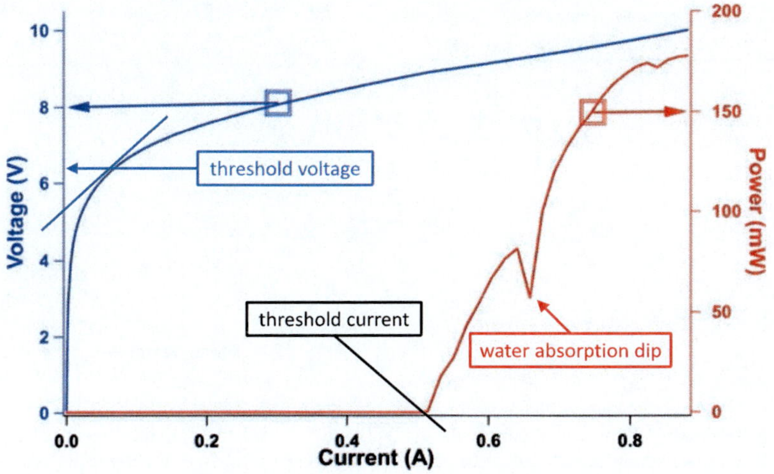

Fig. 2.7 Typical light-current curve and voltage-current curve of one of the DFB-QCLs under study when running CW. The threshold current is 530 mA while the threshold voltage is 6.35 V. The maximum output power is 180 mW and it is possible to observe an absorption dip due to water vapour in the red curve

2.2 Linewidth Enhancement Factor

2.2.1 Definitions

The linewidth enhancement factor (LEF) is a parameter of paramount importance for studying the non-linear dynamics of QCLs that we will develop in the coming chapters. The LEF quantifies the coupling between the gain and the refractive index of the QCL or, in a similar manner, the coupling between the phase and the amplitude of the electrical field [46]. This parameter quantifying the dynamics of semiconductor lasers can also be described as the ratio of the derivatives of the real and imaginary parts of the susceptibility χ with respect to the carrier density n [47]:

$$\alpha = -\frac{\frac{\partial \chi_r}{\partial n}}{\frac{\partial \chi_i}{\partial n}} \tag{2.10}$$

Prior efforts focused on experimental studies of the LEF for pump currents above threshold but without exceeding 12% of the threshold current at 283 K [48] and 56% of the threshold current at 82 K [49], as shown in Fig. 2.8. Consequently, a broader study was required in order to retrieve the LEF for injection currents below threshold and up to the maximum injection current the QCL can endure. Depending on whether the injection current is above or below threshold, two different methods were considered: the amplified spontaneous emission (ASE) method [50] to retrieve the LEF for

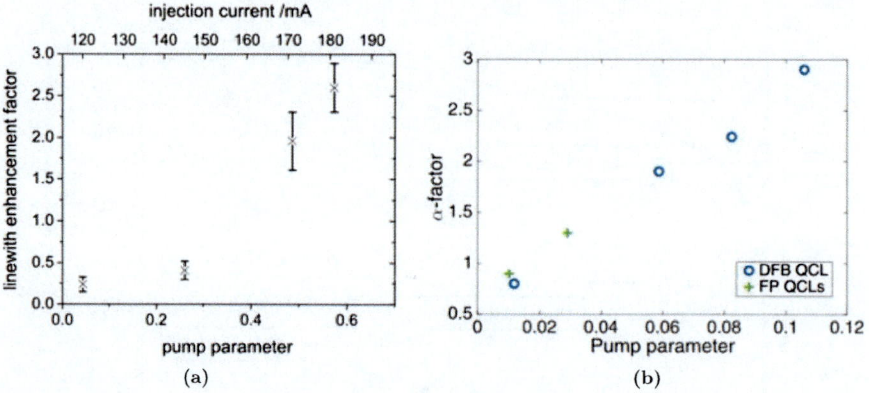

Fig. 2.8 Prior experimental measurements of QCLs' LEF with the self-mixing interferometry technique; **a** is performed at cryogenic temperature and **b** is performed at room temperature but both measurements only focus on the LEF above current threshold. Courtesy of [48, 51]

current biases below threshold, and the self-mixing interferometry technique [51] to obtain LEFs for current biases up to more than twice the threshold current.

The LEF can also be found in the literature under the designation linewidth broadening factor or α-factor, since it is responsible for the linewidth broadening in the optical domain, resulting in a linewidth enhanced by a factor $(1 + \alpha^2)$ compared to the Shawlow-Townes limit [52]. The linewidth of a semiconductor laser thus is [53]:

$$\delta\nu = \frac{v_g^2\, h\,\nu\,\alpha_{tot}\,\alpha_M\,n_{sp}}{4\,\pi\,P_0}\left(1+\alpha^2\right) \tag{2.11}$$

where v_g, n_{sp}, α_{tot}, α_M and P_0 are the group velocity, the spontaneous emission factor (or population inversion parameter), the total cavity losses, the mirror losses and the output power, respectively. If we define the FWHM of the cavity resonance as $\delta\nu_c = \frac{v_g\,\alpha_{tot}}{2\pi}$ and if we assume a lorentzian lineshape with a FWHM equal to 2γ for the intersubband transition, then the linewidth enhancement factor of a QCL can be expressed as a function of the center frequency of the gain spectrum ν_0 [39]:

$$\alpha(\nu) = \frac{\nu_0 - \nu}{\gamma} \tag{2.12}$$

and is thus close to zero because of an homogeneous gain medium. In laser diodes, experiments with several techniques, such as ASE method (sometimes called Hakki-Paoli method) [54], radio-frequency measurements [55], analysis of the locking regimes induced by optical injection from a master laser [56], or optical feedback self-mixing effects [57], have proved that the LEF can take values between 1 and 8. This is known to be caused by the symmetry breaking in the two bands involved in the laser transition, resulting in a spectrally asymmetric differential gain. In contrast, both laser subbands of a QCL are within the conduction band and exhibit the same

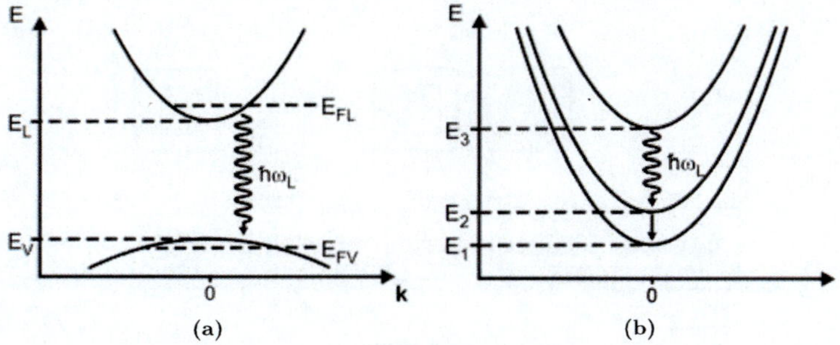

Fig. 2.9 **a** Transition mechanism from conduction to valence band in an interband semiconductor laser. **b** Transition mechanism between two subbands in an intersubband semiconductor laser, e.g. a QCL. Courtesy of [51]

reciprocal space curvature. The difference between these two configurations can be visualized in Fig. 2.9. This should lead to a zero LEF in the case of QCLs, because of a symmetric differential gain. However, experimental efforts in the terahertz [58] and the mid-infrared domain [59], showed that QCLs may exhibit a non-zero LEF.

2.2.2 *Self-mixing Interferometry*

In the case of self-mixing interferometry, the feedback strength must be in the order of 10^{-6} at maximum [60] because otherwise, the self-mixing pattern is combined with non-linear dynamics which will be thoroughly described hereafter. Another condition of paramount importance is that the laser must be a mono-mode laser. Therefore, this method can be applied to DFB-QCLs but not to FP-QCLs. A schematic of the experimental setup allowing retrieving the LEF with this technique can be found in Fig. 2.10. The QCL's beam is split with a non-polarizing beam-splitter and one part is retrieved with a MCT-detector. The other part passes the beam-splitter and is fed back to the laser via a mirror, which is mounted on a linear piezo-translation stage. A neutral-density filter (NDF) in the feedback arm allows controlling the feedback strength and remaining in the weak-feedback regime. The translation stage is driven with a sine modulation of 137 Hz.

The self-mixing interferometry method is a well known technique to obtain the LEF in the case of single-mode semiconductor lasers and has been successfully adapted for DFB-QCLs in the past. The method is based on the dependence of the intracavity mixing signal on the phase difference ϕ between the emitted and the back-reflected light. The so-called feedback phase ϕ can be, for example, varied by modulating the external-cavity length. An interferogram, the so-called selfmixing signal $P(\phi)$, is then obtained as a function of ϕ. Consequently, the optical output power of the laser is given by:

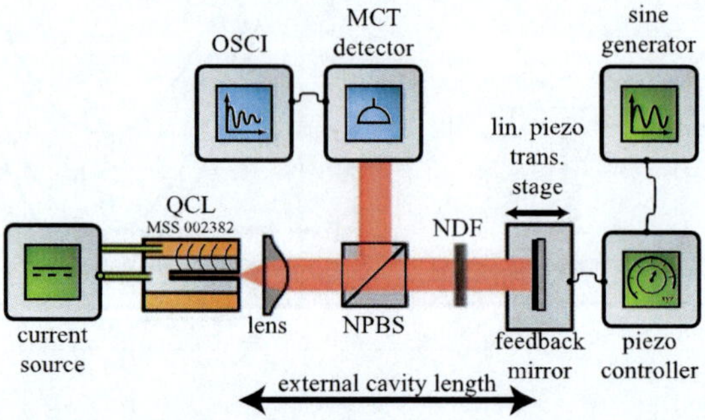

Fig. 2.10 Schematic of the self-mixing interferometry setup with oscilloscope (OSCI), MCT (Mercury-Cadmium-Telluride) detector, non-polarizing beam-splitter (NPBS), low-noise current source, linear piezo-translation stage, piezo controller and connected sine generator, as well as a neutral density filter (NDF) to control the feedback strength

$$P(\phi) = P_0\big(1 + mF(\phi)\big), \tag{2.13}$$

where P_0 represents the laser power without optical feedback and m is a modulation index. The interferometric function $F(\phi)$ depends on the LEF and the feedback parameter C given by:

$$C = \epsilon_{\mathrm{mm}} \frac{L\sqrt{1+\alpha^2}}{l \cdot n} \sqrt{R_{\mathrm{ext}}} \cdot \frac{1-R_1}{\sqrt{R_1}}, \tag{2.14}$$

where ϵ_{mm} represents the mode-mismatch parameter, L the length of the external cavity and l the length of the internal cavity. The variable n is the internal cavity refractive index, while R_{ext} represents the reflectivity of the feedback mirror and R_1 the reflectivity of the output facet. In the case of weak feedback [61] (i.e. $0 < C < 1$), it is possible to express the LEF as:

$$\alpha = \frac{T_{\mathrm{M}} - 0.5}{T_{\mathrm{C}} - 0.5} \tag{2.15}$$

where T_{M} and T_{C} are time-intervals [49], which are defined as:

$$T_{\mathrm{M}} \equiv \frac{T_{\max} - T_{\min}}{T} \tag{2.16}$$

$$T_{\mathrm{C}} \equiv \frac{T_{\mathrm{C2}} - T_{\mathrm{C1}}}{T}. \tag{2.17}$$

The time-points T_{C2} and T_{C1} represent the positions of the consecutive zeros, $T_{\max}$ and $T_{\min}$ the positions of the consecutive extrema and the interval T is the period of the interferogram. In Fig. 2.11a, a cutout of the normalized waveform

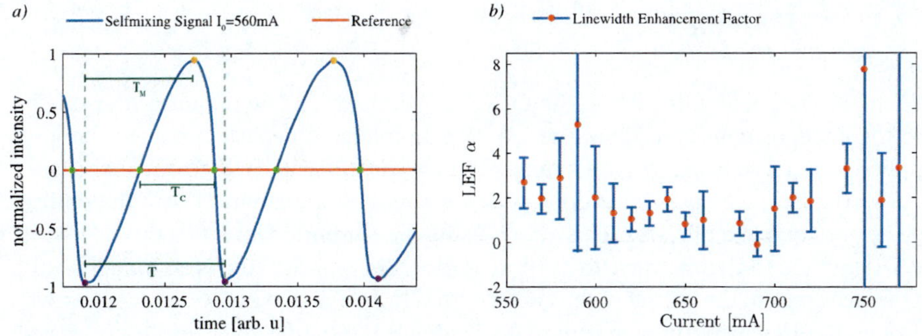

Fig. 2.11 **a** Cutout of the experimentally obtained normalized waveform from the weak-feedback selfmixing signal of the QCL driven with an injection current of $J_0 = 560\,\text{mA}$. The time-intervals T_M, T_C and T required for the determination of the LEF are materialized. The resulting LEF amounts to $\alpha = 2.67 \pm 0.45$. **b** LEF as a function of the injection current

$P_{\text{norm}}(t) = \frac{(P(t)-P_0)}{P_{\text{max}}}$ for an injection current of $J_0 = 560\,\text{mA}$ is exhibited, where $P_{\text{max}} = \max\left(P(t) - P_0\right)$. In that cutout, the mirror moves linearly towards the laser, thus decreasing the feedback phase ϕ linearly. By determining the required time-intervals T_{M}, T_{C} and T, which are highlighted in dark green, together with using Eqs. 2.15–2.17, it is possible to obtain the LEF. The result in the case of an injection current of 560 mA amounts to 2.67 ± 0.45. We do not show LEFs for currents lower than 560 mA, because the self-mixing signal was too noisy in this region, resulting in a very high uncertainty of the LEF value. Figure 2.11b depicts the influence of the injection current on the LEF. For that purpose, 100 waveforms were acquired at each injection current in order to derive the LEF. The orange dots represent the mean values of the determined LEFs, and the blue error-bars depict the standard-deviation of all determinations. It can be seen that for all injection currents, the LEF value remains positive. For low injection currents the LEF is around 2, while it becomes close to 0 for injection currents between 650 mA and 700 mA. For higher injection currents, the LEF increases to values around 4. Such large values could be explained by the pumping far above threshold current but could also result from the DFB grating which alters the structure of the initial FP-laser. This modification can be responsible for spatial non-linearities such as spatial hole-burning (SHB) and can have an influence on the LEF of the QCL [62]. These values high above zero are compatible with what will be described in terms of bifurcation diagrams, both numerically and experimentally. On the one hand, if the LEF is too close to zero, the non-linear dynamics cannot be exhibited, even for large feedback strengths [63] but on the other hand, the intersubband structure prevents LEF values as large as the ones found in laser diodes [64].

2.2.3 ASE Method

When pumped below threshold, the QCL only emits very few output power due to the spontaneous emission. Thus, the self-mixing interferometry technique cannot be applied to retrieve the LEF in this range of injection currents. Moreover, as the laser is powered below threshold, it does not emit single-mode anymore and the optical spectrum is that of a FP-laser. This allows using the Amplified Spontaneous Emission (ASE) method [54], to retrieve the LEF. The aforementioned setup is no longer useful in this section and the output of the QCL is directly analyzed with a Fourier transform infrared (FTIR) spectrometer. Figure 2.12a shows the Fabry-Perot spectra retrieved from 255 to 325 mA. 255 mA is the minimum current value for which a signal can be detected and 325 mA corresponds to the current value just below threshold. The ASE method relies on the evolution of the Fabry-Perot spectra when varying the pump current, and more precisely the shifting of the modes. Figure 2.12b shows a close-up on the modes we will focus on to derive the LEF below threshold. The modal gain G, expressed for a pumping current, is given by [65]:

$$G = \frac{1}{l} ln\Big(\frac{\sqrt{\Delta I} - 1}{\sqrt{\Delta I} + 1}\Big) + \frac{1}{2l} ln\Big(\frac{1}{R_1 R_2}\Big), \tag{2.18}$$

with l the length of the laser's cavity, $\Delta I = I_{\max}/I_{\min}$ the ratio of the maximum and the minimum of the Fabry-Perot spectrum and R_1 and R_2 the reflectivities of the laser's facets.

The change in refractive index Δn is then obtained from the peak-wavelength shift $\Delta\lambda$ when increasing the pumping current [65]:

$$\Delta n = \frac{\lambda}{2l} \times \frac{\Delta\lambda}{\lambda_{\text{FP}}}, \tag{2.19}$$

(a) (b)

Fig. 2.12 Experimental optical spectra below threshold for a pumping current varying from 255 to 325 mA in steps of 5 mA (**a**) and close-up around 5.39 μm (**b**) for pumping currents shown in Fig. 2.13a

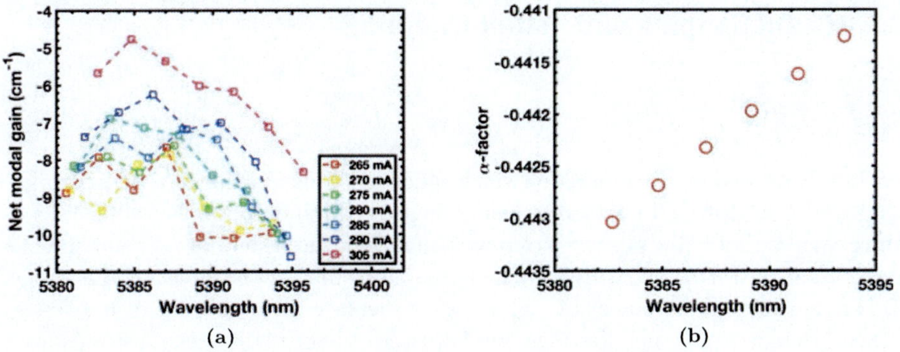

Fig. 2.13 Net modal gain with respect to the wavelength for several pumping current below threshold (**a**) and calculated LEF when comparing the data at 290 and 305 mA (**b**)

with λ_{FP} the mode spacing in the Fabry-Perot spectrum. When increasing the pumping current, the change in the modal gain and the variation of the refractive index are used to extract the LEF [66]:

$$\alpha = -\frac{4\pi}{\lambda} \times \frac{\Delta n}{\Delta G}, \tag{2.20}$$

For the calculation of the LEF, we focus on the optical spectra at 290 mA and 305 mA, whose maxima are marked with green circles and purple circles in Fig. 2.12b, respectively. The experimental data selected for the derivation are those around 5390 nm because the modal gain is more homogeneous in this spectral window (see Fig. 2.12a) than close to the maximum output of the laser, which is fixed at 5450 nm by the DFB grating. The extracted gain spectra are represented in Fig. 2.13a for various pumping current conditions. Since the measurement is done under CW conditions, the wavelength red-shift caused by thermal effects must be fully eliminated so that we only account for the net carrier induced effects [67]. In such purpose, the wavelength shift below threshold is estimated from the wavelength red-shift due to thermal effects above threshold, the latter being 0.49 Å.mA^{-1} between 360 and 500 mA. The spectral dependence of the LEF is plotted in Fig. 2.13b. The retrieved LEF is around −0.4 and differs from the value found with the self-mixing interferometry technique above threshold. The latter is however retrieved for current values high above threshold and cannot be compared with the value found close to and below threshold. Furthermore, numerical simulations have already demonstrated a LEF increasing from negative values to positive values in the case of QCLs [68].

2.3 Beam Shaping and Talbot Coupling

2.3.1 Beam Steering

We briefly digress in this subsection to introduce methods to shape the beam of large QCLs or arrays of QCLs aimed at generating very high output power. Indeed, we already talked about the necessity of powerful mid-infrared sources for countermeasure systems and selective surgery. The two options that can be envisioned in order to increase the optical power of QCLs are an increase of the doping of the active area or a wider semiconductor ridge. First option could be realized because the maximum injection current a QCL can endure is proportional to the number of carriers [69]. However, an increase of the doping would lead to a detrimental increase of the absorption losses and thermal backfilling. The second option deals with broad-area (BA) QCLs. They are envisioned to emit more than 20 W of optical power when running CW [70] but they are concerned with both thermal and optical issues affecting overall beam quality performances. Indeed, even though the thermal resistance decreases with the ridge width, the thermal load becomes too important to be dissipated efficiently. This means that the only option not to detrimentally degrade the performances of the BA-QCL, is to operate it with very short pulses, but this shrinks the mean optical power. Even if the BA-QCL is optimized in terms of thermal properties, thanks to diamond substrate for instance [71], the main phenomenon degrading the beam quality is the beam steering effect. It originates from the large third-order susceptibility $\chi^{(3)}$ in QCLs [72], which leads to a non-linear coupling between the transverse modes and hence, to four-wave mixing interaction and phase-coherence [73]. The electric field in the cavity can be expressed as a combination of the fields of all the transverse modes, resulting in a deteriorated far-field [74]. This beam steering effect appears as soon as more than one transverse mode co-exist in the cavity, i.e. for active regions typically broader than 12 μm for a QCL emitting around 5 μm . In order to address the beam quality of BA-QCLs, several solutions have been proposed. QCLs with photonic crystals etched on top of the active region [75, 76] or with a tilted facet [77] have led to an improved diffraction-limited single-lobe far-field. Also, even if they present a smaller gain region compared to BA-QCLs, tapered QCLs are a relevant compromise between large effective active region, high output power and upper beam quality [78]. Another approach to solve both thermal and optical drawbacks of BA-QCLs is to split the ridge into an array of micro-stripes optically coupled to each other to achieve a stable optical supermode. The far-field is typically two-lobed in the case of evanescent coupling [79], but single-lobe emission has been achieved in the case where the stripes are coupled by antiguiding [80].

All the aforementioned processes rely on monolithic integration and consequently depend on the fabrication steps repeatability and quality. Other methods have been explored in order to shape the multi-mode beacon of BA-lasers or arrays of lasers, as well as for reducing filamentation [81]. Filamentation corresponds to fast spatio-temporal oscillations due to diffraction of light, self-focusing and spatial hole-burning [82], whose position along the laser cavity fluctuates with time. Even though the origin of filamentation is not fully understood yet, it leads to the excitation of higher spatial modes, with different phase velocities, hence deteriorating significantly the laser coherence and the beam quality. Moreover, the appearance of filamentation in a BA-laser is related to the linewidth enhancement factor (LEF) value of the device [83]. The higher the LEF, the more the laser beam quality will be impacted by filamentation. Optical feedback can be used to counter the filamentation-induced drawbacks, without altering other performances of the laser. Most of the non-integrated techniques tackling filamentation and beam steering take advantage of reinjection [84–86]. The impact of optical feedback on the near-field profile of a BA semiconductor laser is ruled by three main parameters. The first one is the sign of the population-inversion induced index change, i.e. whether the laser design is based on gain-guiding (positive index variation) or index-guiding (negative index variation). Studies have shown that total stabilization of the emission pattern, where optical feedback forces the laser to operate on the fundamental transverse mode, can only be achieved in the case of negative population-induced index change [87]. Furthermore, the two other key parameters are the feedback conditions: depending on the feedback ratio and the external-cavity length, higher spatial modes will be either excited or suppressed [85]. Feedback ratio defines the ratio between reinjected and emitted powers. Spatially-filtered optical feedback can further improve the near-field profile of the laser emission. Reinjecting only the central part of the emitted beam will indeed favor the lower order modes, leading to a high-quality beam profile close to the single-transverse mode behavior [85]. In the case of a 14 μm-wide and 3 mm-long QCL emitting around 4.6 μm, simulations showed that two transverse modes exist in the cavity, as illustrated in Fig. 2.14. Jumpertz and co-workers were able to obtain a gaussian near-field profile with a centered optical feedback [88], contrasting with the initial flat-topped pattern shown in Fig. 2.14b. The method was proven efficient for even wider QCLs [89]. In the case of a 40 μm-wide and 3 mm-long QCL emitting around 4.7 μm, 9 modes can be observed in the near-field of Fig. 2.15a without back-injection, although the intensity of each antimode varies and the furthest to the right is quite dim. Applying optical feedback, as shown in Fig. 2.15b, favors one dominant peak in the near-field. Though it is not the fundamental mode of the BA-QCL, it shows that distinct modes can be enhanced with optical feedback. The light-current curves in Fig. 2.15c show that the feedback increases the total power emitted at the facet, for a given bias current. So not only the beam divergence is reduced, but also the output power is optimized, even if it is not possible to totally suppress the side-modes, as explained in the case of the 14 μm-wide QCL.

2.3.2 *Talbot Coupling in a Laser Array*

Another method to combine low divergence and high power relies on Talbot coupling. It is not possible to apply it on a BA-laser but it was proven useful in the case of an array of semiconductor lasers emitting on both facets. Talbot effect is a well-known lensless optical phenomenon that the intensity pattern of an array of coherent

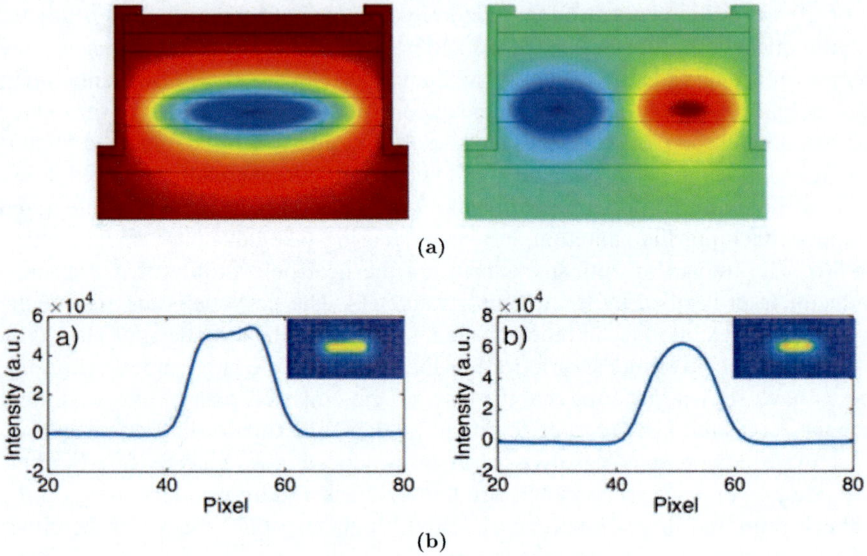

Fig. 2.14 **a** Simulation of the two transverse modes which can exist in the cavity for a 14 μm-wide QCL emitting at 4.6 μm. **b** Near-field profiles of the same QCL without feedback (left) and under centered optical feedback (right). Courtesy of [88]

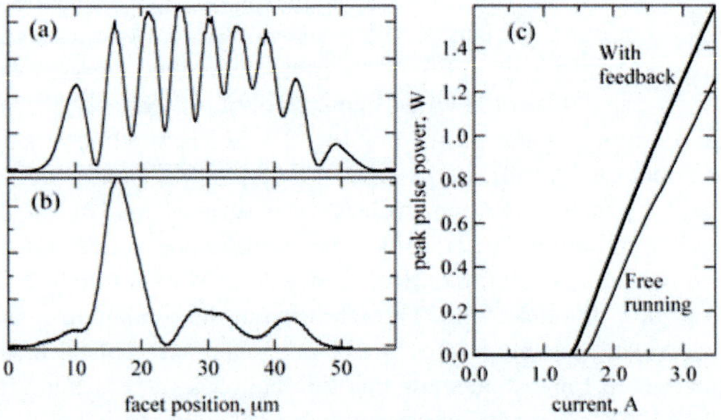

Fig. 2.15 **a** Line scan from near-field profile of a free-running 40 μm-wide QCL, **b** same QCL under optical feedback, forcing single-peak mode. **c** Light-current curves for the free-running and the single-peak feedback case. Courtesy of [89]

emitters reproduces itself after a specific distance of propagation [90, 91]. This effect has been exploited to phase-lock lasers in the near-infrared, which is called diffraction-coupling scheme phase-locked array [92]. If the Talbot effect is applied to QCLs, the setup consists in an array of N QCLs with the same ridge length and placed side by side with a center-to-center distance d. The facets of the QCLs are cleaved on both sides so that emission occurs on both sides. On one side, the global

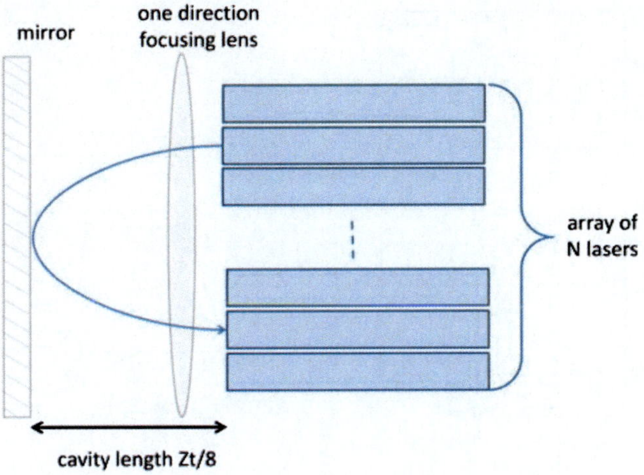

Fig. 2.16 Schematic of the experimental Talbot setup when an array of N lasers is considered. In that case, the length of the Talbot cavity is set to $\frac{Z_t}{8}$ and therefore the image at $\frac{Z_t}{4}$ is sent back to the array

emission properties of the array can be studied with that specific configuration. On the other side, a mirror is placed at an integer multiple of $\frac{Z_t}{8}$, with Z_t being the Talbot distance and scaling as $\frac{2nd^2}{\lambda}$; λ is the emission wavelength of the QCL and n is the refractive index of the medium between the QCLs array and the mirror. For instance, if the mirror is set at a distance equal to $\frac{Z_t}{8}$ as shown in Fig. 2.16, the pattern derived from the Talbot effect that is found at a distance $\frac{Z_t}{4}$ is back reflected to the array. This induces a coupling which can improve the emission properties of the linear QCL array, in particular for creating a bright infrared source with Watt-level power. Experimental efforts have assessed the efficiency of integrated Talbot cavities in the case of an array of QCLs [93–95]. As shown in Fig. 2.17a, it was possible to achieve a powerful beam with a narrow central mode when studying the far-field of an array of 6 QCLs with a 11.5 μm center-to-center spacing. The central mode is almost suppressed in the far-field of a BA-QCL, as illustrated in Fig. 2.17b, so the Talbot coupling allows keeping optical power within a low-divergence cone. Furthermore, the Talbot-cavity QCL emits an optical power that scales almost linearly with the number of devices of the array, as can be seen in Fig. 2.18.

A numerical non-linear analysis of the Talbot effect was recently performed and showed that the array of QCLs can also be prone to non-linear phenomena due to the strong optical feedback induced by the external-Talbot cavity. These non-linear dynamics arise from tiny differences in the size of the lasers' ridge and may be caused by an uneven heat transfer throughout the array. This leads to the possibility that all the lasers will not operate on the same frequency, thus introducing detuning to the array. The consequence is that the QCLs may oscillate at a small frequency, but the numerical calculation indicates that the frequencies of the lasers are not locked

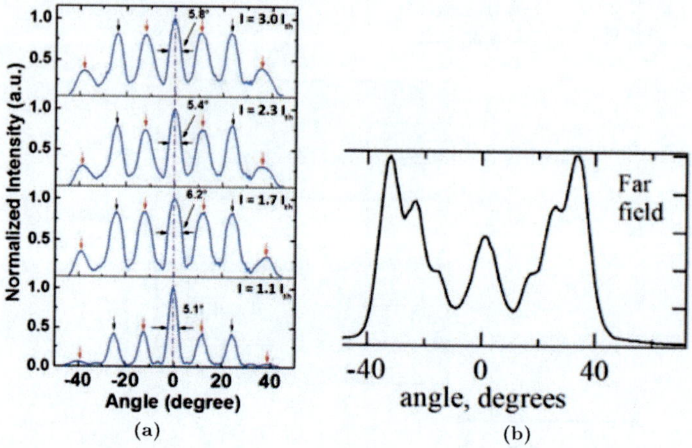

Fig. 2.17 **a** Far-field distribution of the six-laser Talbot-cavity QCL at several pumping levels, showing that part of the output power is gathered in a centered cone. **b** Far-field profile of a 40 μm-QCL with most of the output power in high-divergence side-lobes. Courtesy of [89, 94]

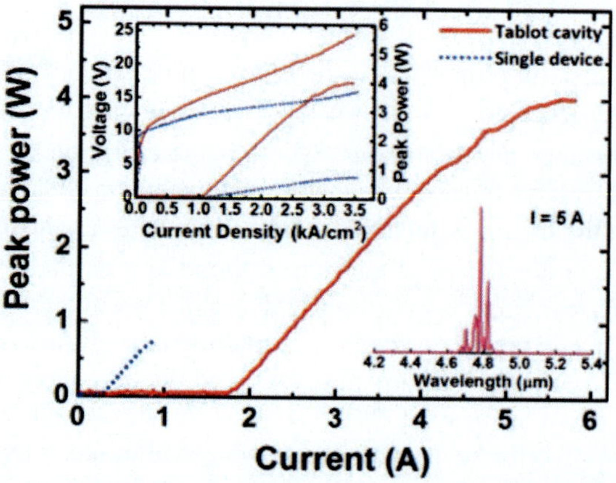

Fig. 2.18 Light-current curves for the six-laser Talbot-cavity QCL and a reference FP-QCL at 300 K. The upper inset shows the light-current curve and the voltage-current curve for the same devices, plotted on a current-density scale. The lower inset shows the optical spectrum of the six-laser Talbot-cavity QCL at a pumping current of 5 A. Courtesy of [94]

and not related to each other. The same conclusion is true for the amplitudes of the lasers. However, the behavior changes with the global feedback strength, and stable emission as well as mode-locked fluctuations have also been observed, specifically for high feedback strengths. This could be of particular interest in mode-locking applications, especially when the number of lasers in a bar becomes rather large.

2.3.3 Conclusion

We described the state-of-the-art of QCLs which are versatile optical sources able to emit in the mid-infrared domain or in the terahertz domain. Thanks to improvements in the growing process and the heat dissipation process, QCLs can be pumped CW and generate up to several Watts of optical power. Recent insights about QCLs on silicon could widespread this technology in integrated photonics systems. The next parts of the dissertation will focus on DFB-QCLs from mirSense emitting around 5.6 μm and with only one transverse mode thanks to the narrow width of the semiconductor ridge.

QCLs with broader ridges were shown to exhibit distorted near-field and far-field profiles even though they are renown for emitting more optical power. An appropriate optical feedback can gather most of the output light into a low-divergence cone while improving the output power of the BA-QCL. In order to achieve powerful mid-infrared beams, it is also possible to build an array of QCLs emitting on both facets. One side is fed back to the array in order to create a Talbot cavity that allows combining the beams in a narrower cone. We consequently saw two advantages of optical feedback in QCLs but this only concerned the laser's intensity at a very long timescale. The next chapters will focus on the dynamical behavior of QCLs subject to external optical feedback, that occurring at a faster timescale.

References

1. Kazarinov R (1971) Possibility of amplification of electromagnetic waves in a semiconductor with superlattice. Sov Phys Semicond 5(4):707–709
2. Tsu R, Esaki L (1973) Tunneling in a finite superlattice. Appl Phys Lett 22(11):562–564
3. Chang L, Esaki L, Tsu R (1974) Resonant tunneling in semiconductor double barriers. Appl Phys Lett 24(12):593–595
4. Esaki L, Chang L (1974) New transport phenomenon in a semiconductor "superlattice". Phys Rev Lett 33(8):495
5. Dingle R, Wiegmann W, Henry CH (1974) Quantum states of confined carriers in very thin $Al_xGa_{1-x}As - GaAs - Al_xGa_{1-x}As$ heterostructures. Phys Rev Lett 33(14):827
6. Cho AY, Arthur J (1975) Molecular beam epitaxy. Progr Solid State Chem 10:157–191
7. Capasso F (1987) Band-gap engineering: from physics and materials to new semiconductor devices. Science 235(4785):172–176
8. Faist J, Capasso F, Sivco DL, Sirtori C, Hutchinson AL, Cho AY (1994) Quantum cascade laser. Science 264(5158):553–556
9. Faist J, Capasso F, Sirtori C, Sivco DL, Hutchinson AL, Cho AY (1995) Continuous wave operation of a vertical transition quantum cascade laser above T = 80 K. Appl Phys Lett 67(21):3057–3059
10. Faist J, Capasso F, Sirtori C, Sivco DL, Baillargeon JN, Hutchinson AL, Chu S-NG, Cho AY (1996) High power mid-infrared ($\lambda \approx 5$ μm) quantum cascade lasers operating above room temperature. Appl Phys Lett 68(26):3680–3682
11. Faist J, Gmachl C, Capasso F, Sirtori C, Sivco DL, Baillargeon JN, Cho AY (1997) Distributed feedback quantum cascade lasers. Appl Phys Lett 70(20):2670–2672
12. Faist J, Gmachl C, Striccoli M, Sirtori C, Capasso F, Sivco DL, Cho AY (1996) Quantum cascade disk lasers. Appl Phys Lett 69(17):2456–2458

13. Gmachl C, Capasso F, Narimanov E, Nöckel JU, Stone AD, Faist J, Sivco DL, Cho AY (1998) High-power directional emission from microlasers with chaotic resonators. Science 280(5369):1556–1564
14. Scamarcio G, Capasso F, Sirtori C, Faist J, Hutchinson AL, Sivco DL, Cho AY (1997) High-power infrared (8-micrometer wavelength) superlattice lasers. Science 276(5313):773–776
15. Schrenk W, Finger N, Gianordoli S, Hvozdara L, Strasser G, Gornik E (2000) Surface-emitting distributed feedback quantum-cascade lasers. Appl Phys Lett 77(14):2086–2088
16. Evans A, Yu J, David J, Doris L, Mi K, Slivken S, Razeghi M (2004) High-temperature, high-power, continuous-wave operation of buried heterostructure quantum-cascade lasers. Appl Phys Lett 84(3):314–316
17. Devenson J, Barate D, Cathabard O, Teissier R, Baranov A (2006) Very short wavelength (λ= 3.1–3.3 μ m) quantum cascade lasers. Appl Phys Lett 89(19):91115
18. Lyakh A, Zory P, D'Souza M, Botez D, Bour D (2007) Substrate-emitting, distributed feedback quantum cascade lasers. Appl Phys Lett 91(18):181116
19. Spott A, Peters J, Davenport ML, Stanton EJ, Merritt CD, Bewley WW, Vurgaftman I, Kim CS, Meyer JR, Kirch J et al (2016) Quantum cascade laser on silicon. Optica 3(5):545–551
20. Spott A, Stanton EJ, Volet N, Peters JD, Meyer JR, Bowers JE (2017) Heterogeneous integration for mid-infrared silicon photonics. IEEE J Select Topics Quant Electron 23(6):1–10
21. Zhuo N, Zhang J-C, Wang F-J, Liu Y-H, Zhai S-Q, Zhao Y, Wang D-B, Jia Z-W, Zhou Y-H, Wang L-J et al (2017) Room temperature continuous wave quantum dot cascade laser emitting at 7.2 μm. Opt Exp 25(12):13 807–13 815
22. Benisty H, Sotomayor-Torres C, Weisbuch C (1991) Intrinsic mechanism for the poor luminescence properties of quantum-box systems. Phys Rev B 44(19):10945
23. Bandyopadhyay N, Slivken S, Bai Y, Razeghi M (2012) High power, continuous wave, room temperature operation of $\lambda \approx 3.4$ μm and $\lambda \approx 3.55$ μm InP-based quantum cascade lasers. Appl Phys Lett 100(21):212104
24. Lyakh A, Maulini R, Tsekoun A, Go R, Von der Porten S, Pflügl C, Diehl L, Capasso F, Patel CK (2010) High-performance continuous-wave room temperature 4.0-μm quantum cascade lasers with single-facet optical emission exceeding 2 W. In: Proceedings of the national academy of sciences, vol 107, no 44, pp 18 799–18 802
25. Bai Y, Bandyopadhyay N, Tsao S, Slivken S, Razeghi M (2011) Room temperature quantum cascade lasers with 27% wall plug efficiency. Appl Phys Lett 98(18):181102
26. Maulini R, Lyakh A, Tsekoun A, Patel CKN (2011) $\lambda \approx 7.1$ μm quantum cascade lasers with 19% wall-plug efficiency at room temperature. Opt Expr 19(18):17 203–17 211
27. Diehl L, Bour D, Corzine S, Zhu J, Höfler G, Lončar M, Troccoli M, Capasso F (2006) High-power quantum cascade lasers grown by low-pressure metal organic vapor-phase epitaxy operating in continuous wave above 400 K. CKN 88(20):201115
28. Slivken S, Evans A, Zhang W, Razeghi M (2007) High-power, continuous-operation intersubband laser for wavelengths greater than 10 μm. Appl Phys Lett 90(15):151115
29. Colombelli R, Capasso F, Gmachl C, Hutchinson AL, Sivco DL, Tredicucci A, Wanke MC, Sergent AM, Cho AY (2001) Far-infrared surface-plasmon quantum-cascade lasers at 21.5 μm and 24 μm wavelengths. Appl Phys Lett 78(18):2620–2622
30. Köhler R, Tredicucci A, Beltram F, Beere HE, Linfield EH, Davies AG, Ritchie DA, Iotti RC, Rossi F (2002) Terahertz semiconductor-heterostructure laser. Nature 417(6885):156
31. Wienold M, Röben B, Schrottke L, Sharma R, Tahraoui A, Biermann K, Grahn H (2014) High-temperature, continuous-wave operation of terahertz quantum-cascade lasers with metal-metal waveguides and third-order distributed feedback. Opt Exp 22(3):3334–3348
32. Wade A, Fedorov G, Smirnov D, Kumar S, Williams BS, Hu Q, Reno JL (2009) Magnetic-field-assisted terahertz quantum cascade laser operating up to 225 K. Nat Photon 3(1):41
33. Lu Q, Bandyopadhyay N, Slivken S, Bai Y, Razeghi M (2012) Widely tuned room temperature terahertz quantum cascade laser sources based on difference-frequency generation. Appl Phys Lett 101(25):251121
34. Davies AG, Burnett AD, Fan W, Linfield EH, Cunningham JE (2008) Terahertz spectroscopy of explosives and drugs. Mater Today 11(3):18–26

35. Amanti MI, Scalari G, Beck M, Faist J (2012) Stand-alone system for high-resolution, real-time terahertz imaging. Opt Exp 20(3):2772–2778
36. Vitiello MS, Scalari G, Williams BS, De Natale P (2015) Quantum cascade lasers: 20 years of challenges. Opt Exp 23(4):5167–5182
37. Yao Y, Hoffman AJ, Gmachl CF (2012) Mid-infrared quantum cascade lasers. Nat Photon 6(7):432
38. Gensty T, Elsäßer W (2005) Semiclassical model for the relative intensity noise of intersubband quantum cascade lasers. Opt Commun 256(1–3):171–183
39. Faist J (2013) Quantum cascade lasers. OUP Oxford
40. Ferré S (2016) Nouvelles architectures et optimisations pour la montée en puissance des lasers à cascade quantique moyen infrarouge, PhD dissertation, École Normale Supérieure
41. Toor F, Sivco DL, Liu HE, Gmachl CF (2008) Effect of waveguide sidewall roughness on the threshold current density and slope efficiency of quantum cascade lasers. Appl Phys Lett 93(3):031104
42. Bai Y, Slivken S, Darvish S, Razeghi M (2008) Room temperature continuous wave operation of quantum cascade lasers with 12.5% wall plug efficiency. Appl Phys Lett 93(2):021103
43. Bai Y, Bandyopadhyay N, Tsao S, Selcuk E, Slivken S, Razeghi M (2010) Highly temperature insensitive quantum cascade lasers. Appl Phys Lett 97(25):251104
44. Carras M, Garcia M, Marcadet X, Parillaud O, De Rossi A, Bansropun S (2008) Top grating index-coupled distributed feedback quantum cascade lasers. Appl Phys Lett 93(1):011109
45. Carras M, Maisons G, Simozrag B, Garcia M, Parillaud O, Massies J, Marcadet X (2010) Room-temperature continuous-wave metal grating distributed feedback quantum cascade lasers. Appl Phys Lett 96(16):161105
46. Henry C (1982) Theory of the linewidth of semiconductor lasers. IEEE J Quant Electron 18(2):259–264
47. Haken H (1970) Laser theory. In: Light and Matter Ic/Licht und Materie Ic. Springer, pp 1–304
48. Jumpertz L, Michel F, Pawlus R, Elsässer W, Schires K, Carras M, Grillot F (2016) Measurements of the linewidth enhancement factor of mid-infrared quantum cascade lasers by different optical feedback techniques. AIP Adv 6(1):015212
49. Von Staden J, Gensty T, Elsäßer W, Giuliani G, Mann C (2006) Measurements of the α factor of a distributed-feedback quantum cascade laser by an optical feedback self-mixing technique. Opt Lett 31(17):2574–2576
50. Hakki BW, Paoli TL (1975) Gain spectra in GaAs double- heterostructure injection lasers. J Appl Phys 46(3):1299–1306
51. Donati S (1978) Laser interferometry by induced modulation of cavity field. J Appl Phys 49(2):495–497
52. Schawlow AL, Townes CH (1958) Infrared and optical masers. Phys Rev 112(6):1940
53. Osinski M, Buus J (1987) Linewidth broadening factor in semiconductor lasers-an overview. IEEE J Quant Electron 23(1):9–29
54. Henning I, Collins J (1983) Measurements of the semiconductor laser linewidth broadening factor. Electron Lett 19(22):927–929
55. Kikuchi K, Okoshi T (1985) Estimation of linewidth enhancement factor of AlGaAs lasers by correlation measurement between FM and AM noises. IEEE J Quant Electron 21(6):669–673
56. Liu G, Jin X, Chuang S-L (2001) Measurement of linewidth enhancement factor of semiconductor lasers using an injection-locking technique. IEEE Photon Technol Lett 13(5):430–432
57. Yu Y, Giuliani G, Donati S (2004) Measurement of the linewidth enhancement factor of semiconductor lasers based on the optical feedback self-mixing effect. IEEE Photon Technol Lett 16(4):990–992
58. Green RP, Xu J-H, Mahler L, Tredicucci A, Beltram F, Giuliani G, Beere HE, Ritchie DA (2008) Linewidth enhancement factor of terahertz quantum cascade lasers. Appl Phys Lett 92(7):071106
59. Hangauer A, Wysocki G (2015) Gain compression and linewidth enhancement factor in mid-IR quantum cascade lasers. IEEE J Select Topics Quant Electron 21(6):74–84

60. Cardilli MC, Dabbicco M, Mezzapesa FP, Scamarcio G (2016) Linewidth measurement of mid infrared quantum cascade laser by optical feedback interferometry. Appl Phys Lett 108(3):031105
61. von Staden J, Gensty T, Peil M, Elsäßer W, Giuliani G, Mann C (2006) Measurement of the linewidth enhancement factor of quantum cascade lasers by the self-mixing technique. In: Semiconductor lasers and laser dynamics II, vol. 6184. International Society for Optics and Photonics, p 61841E
62. Gordon A, Wang CY, Diehl L, Kärtner FX, Belyanin A, Bour D, Corzine S, Höfler G, Liu H, Schneider H et al (2008) Multimode regimes in quantum cascade lasers: from coherent instabilities to spatial hole burning. Phys Rev A 77(5):053804
63. Erneux T, Kovanis V, Gavrielides A (2013) Nonlinear dynamics of an injected quantum cascade laser. Phys Rev E 88(3):032907
64. Kim J, Lerttamrab M, Chuang SL, Gmachl C, Sivco DL, Capasso F, Cho AY (2004) Theoretical and experimental study of optical gain and linewidth enhancement factor of type-I quantum-cascade lasers. IEEE J Quant Electron 40(12):1663–1674
65. Minch J, Chuang S-L, Chang C-S, Fang W, Chen Y-K, Tanbun-Ek T (1997) Theory and experiment on the amplified spontaneous emission from distributed-feedback lasers. IEEE J Quant Electron 33(5):815–823
66. Lerttamrab M, Chuang S, Gmachl C, Sivco D, Capasso F, Cho A (2003) Linewidth enhancement factor of a type-I quantum-cascade laser. J Appl Phys 94(8):5426–5428
67. Wang C, Schires K, Osiński M, Poole PJ, Grillot F (2016) Thermally insensitive determination of the linewidth broadening factor in nanostructured semiconductor lasers using optical injection locking. Sci Rep 6:27825
68. Pereira M (2016) The linewidth enhancement factor of intersubband lasers: From a two-level limit to gain without inversion conditions. Appl Phys Lett 109(22):222102
69. Aellen T, Beck M, Hoyler N, Giovannini M, Faist J, Gini E (2006) Doping in quantum cascade lasers. I. InAlAs–InGaAs/InP midinfrared devices. J Appl Phys 100(4), 043101
70. Suttinger M, Go R, Figueiredo P, Lyakh A (2018) Towards 20-watt continuous wave quantum cascade lasers. In: Micro-and nanotechnology sensors, systems, and applications X, vol 10639. International Society for Optics and Photonics, p 1063922
71. Chaparala SC, Xie F, Caneau C, Zah CE, Hughes LC (2011) Design guidelines for efficient thermal management of mid-infrared quantum cascade lasers. IEEE Trans Compon Packag Manufact Technol 1(12):1975–1982
72. Friedli P, Sigg H, Hinkov B, Hugi A, Riedi S, Beck M, Faist J (2013) Four-wave mixing in a quantum cascade laser amplifier. Appl Phys Lett 102(22):222104
73. Yu N, Diehl L, Cubukcu E, Bour D, Corzine S, Höfler G, Wojcik AK, Crozier KB, Belyanin A, Capasso F (2009) Coherent coupling of multiple transverse modes in quantum cascade lasers. Phys Rev Lett 102(1):013901
74. Bewley WW, Lindle JR, Kim CS, Vurgaftman I, Meyer JR, Evans AJ, Yu JS, Slivken S, Razeghi M (2005) Beam steering in high-power CW quantum-cascade lasers. IEEE J Quant Electron 41(6):833–841
75. Vurgaftman I, Meyer JR (2002) Photonic-crystal distributed-feedback quantum cascade lasers. IEEE J Quant Electron 38(6):592–602
76. Gökden B, Bai Y, Bandyopadhyay N, Slivken S, Razeghi M (2010) Broad area photonic crystal distributed feedback quantum cascade lasers emitting 34 W at $\lambda \approx 4.36\ \mu$m. Appl Phys Lett 97(13):131112
77. Ahn S, Schwarzer C, Zederbauer T, MacFarland DC, Detz H, Andrews AM, Schrenk W, Strasser G (2014) High-power, low-lateral divergence broad area quantum cascade lasers with a tilted front facet. Appl Phys Lett 104(5):051101
78. Lyakh A, Maulini R, Tsekoun A, Go R, Patel CKN (2012) Tapered 4.7 μm quantum cascade lasers with highly strained active region composition delivering over 4.5 watts of continuous wave optical power. Opt Exp 20(4):4382–4388
79. De Naurois G, Carras M, Simozrag B, Patard O, Alexandre F, Marcadet X (2011) Coherent quantum cascade laser micro-stripe arrays. Aip Adv 1(3):032165

80. Kirch JD, Chang CC, Boyle C, Mawst LJ, Lindberg III D, Earles T, Botez D (2015) 5.5 W near-diffraction-limited power from resonant leaky-wave coupled phase-locked arrays of quantum cascade lasers. Appl Phys Lett 106(6):061113
81. Marciante JR, Agrawal GP (1996) Nonlinear mechanisms of filamentation in broad-area semiconductor lasers. IEEE J Quant Electron 32(4):590–596
82. Aoki M, Uomi K, Tsuchiya T, Sasaki S, Chinone N (1990) Stabilization of the longitudinal mode against spatial hole burning in λ/4-shifted DFB lasers by quantum size effect. IEEE Photon Technol Lett 2(9):617–619
83. Dente GC (2001) Low confinement factors for suppressed filaments in semiconductor lasers. IEEE J Quant Electron 37(12):1650–1653
84. Paboeuf D, Lucas-Leclin G, Georges P, Michel N, Krakowski M, Lim J, Sujecki S, Larkins E (2008) Narrow-line coherently combined tapered laser diodes in a Talbot external cavity with a volume Bragg grating. Appl Phys Lett 93(21):211102
85. Takeda A, Shogenji R, Ohtsubo J (2013) Spatial-mode analysis in broad-area semiconductor lasers subjected to optical feedback. Optic Rev 20(4):308–313
86. Ferré S, Jumpertz L, Carras M, Ferreira R, Grillot F (2017) Beam shaping in high-power broad-area quantum cascade lasers using optical feedback. Sci Rep 7:44284
87. Martin-Regalado J, Van Tartwijk G, Balle S, San Miguel M (1996) Mode control and pattern stabilization in broad-area lasers by optical feedback. Phys Rev A 54(6):5386
88. Jumpertz L, Ferré S, Carras M, Grillot F (2017) Beam steering in quantum cascade lasers with optical feedback. In: Quantum sensing and nano electronics and photonics XIV, vol 10111. International Society for Optics and Photonics, p 101112F
89. Grillot F, Newell T, Gavrielides A, Carras M (2017) Complex delay dynamics of high power quantum cascade oscillators. In: Terahertz emitters, receivers, and applications VIII, vol 10383. International Society for Optics and Photonics, p 103830D
90. Talbot HF (1836) Lxxvi. facts relating to optical science. no. iv. London Edinburgh Dublin Philosoph Mag J Sci 9(56):401–407
91. Rayleigh L (1881) Xxv. on copying diffraction-gratings, and on some phenomena connected therewith. London Edinburgh Dublin Philosoph Mag J Sci 11(67):196–205
92. Corcoran CJ, Durville F (2005) Experimental demonstration of a phase-locked laser array using a self-Fourier cavity. Appl Phys Lett 86(20):201118
93. Wang L, Zhang J, Jia Z, Zhao Y, Liu C, Liu Y, Zhai S, Ning Z, Xu X, Liu F (2016) Phase-locked array of quantum cascade lasers with an integrated Talbot cavity. Opt Exp 24(26):30 275–30 281
94. Meng B, Qiang B, Rodriguez E, Hu XN, Liang G, Wang QJ (2017) Coherent emission from integrated Talbot-cavity quantum cascade lasers. Opt Exp 25(4):3077–3082
95. Botez D (2017) Comment on phase-locked array of quantum cascade lasers with an intracavity spatial filter. Appl Phys Lett 111:061108, Appl Phys Lett 111(25):256101

Chapter 3
Chaos in Quantum Cascade Lasers

This chapter will start with a simple theoretical model to introduce chaos in a global manner. Fluctuations observed since the early days of lasers will then be related to chaotic fluctuations and techniques to assess the chaotic nature of the observed fluctuations will be described. As recent experiments in QCLs have shown the basic components of chaotic dynamics, these results will highlight the diversity of such non-linear phenomenon. Parameters having a strong influence on the observed dynamics will be explored, and more precisely the temperature, the carrier-to-photon lifetime and the level of injection current. We will see that QCLs are more prone to chaos at low temperature and that a quasi-continuous bias is more appropriate to generate complex dynamics. In order to explore all the degrees of freedom offered by the external feedback configuration, the current bias will be modulated and this will lead to a synchronization of the chaotic fluctuations with the modulation wave. The evolution of the frequency and phase-shift of the spiking pattern will be reviewed and will give first indications about the use of chaos for communication purposes. The end of the chapter will be dedicated to two other reinjection methods, namely phase-conjugate feedback and cross-polarization reinjection. The latter methods is able to trigger all-optical square-wave patterns whose characteristics will be analyzed, albeit not yet understood with a numerical analysis.

3.1 Chaos Theory

3.1.1 Instabilities in Lasers: From Noise to Chaos

The first instabilities observed in the light generated by lasers [1] were not considered as a chaotic behavior. Pulses with an amplitude varying erratically, as the ones seen in Fig. 3.1, were reported in the output of a ruby solid-state laser and then found in

O. Spitz, *Mid-infrared Quantum Cascade Lasers for Chaos Secure Communications*, Springer Theses, https://doi.org/10.1007/978-3-030-74307-9_3

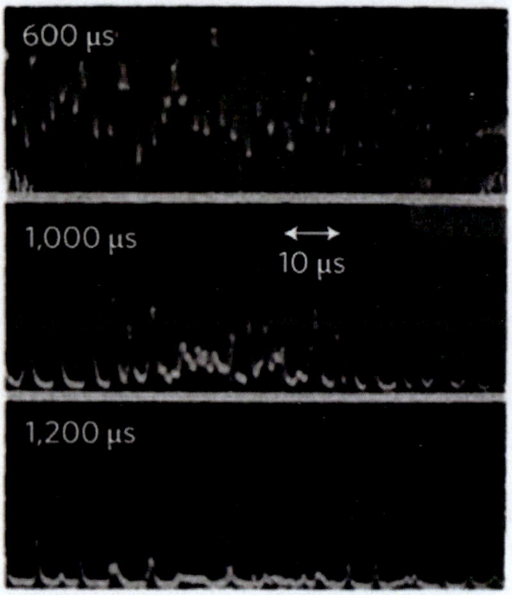

Fig. 3.1 Irregular pulsing from a ruby laser, considered as the first experimental proof of chaotic dynamics in the light emitted by such optical source, even if the concept of chaos was not applied to lasers yet. Courtesy of [1]

numerical simulations [2]. A phase-diffusion model was introduced soon after the laser discovery [3] to explain the broadening of the optical linewidth of lasers. In this model, the absence of spontaneous emission implies that the laser has a negligible linewidth and the related equation is a Langevin equation, which has been thoroughly used in the study of the Brownian motion. In the case of laser dynamics, the set of equations is written [4]:

$$\frac{dE(t)}{dt} = aE(t) - b|E(t)|^2E(t) + f(t) \tag{3.1}$$

$$\langle f(t)^* f(t') \rangle = 2D\delta(t - t') \tag{3.2}$$

with $E(t)$ the complex electric field. The parameter a represents the net gain, while b represents the saturation. The dynamic properties of the field also include a random noise $f(t)$ related to the spontaneous emission fluctuations, with D the strength of that noise. With this set of equations, the amplitude and the phase of the laser can fluctuate with time while the laser is on a steady-state mode. However, some lasers were quickly found to exhibit dynamics which were much more complex than those described with this set of equations, but the first explanations for such behavior were attributed to external-noise sources. The idea that chaos could be involved in the process was then investigated in parallel with the theory used by Lorenz for weather forecast [5]. This model consists of three equations with a few non-linear terms and with no stochastic term. This model has just enough nonlinearity and coupled degrees of freedom to generate chaotic dynamical behaviors. Indeed, the Poincaré

-Bendixson theorem states that at least three degrees of freedom are required to observe deterministic chaos in continuous-time dynamical systems and this was not possible with a Langevin equation [6]. Another peculiarity of the Lorenz model is that it includes a strong sensitivity to initial conditions for the variables and this leads to trajectories diverging exponentially from each other in the phase space. This typical signature of deterministic chaos remains as one of the most acknowledged characteristics of this kind of dynamics. The first model linking the laser behavior with this new dynamics was established in 1975 [7]. The equations of this model were derived from the Maxwell-Bloch equations describing the time evolution of the atoms and the electric field of the laser. An extensive demonstration of this derivation can be found in Ref. [8]. These three equations, called the Lorenz-Haken equations [9], are written:

$$\frac{dx(t)}{dt} = \sigma\big(y(t) - x(t)\big) \tag{3.3}$$

$$\frac{dy(t)}{dt} = -x(t)z(t) + rx(t) - (1 - i\delta)y(t) \tag{3.4}$$

$$\frac{dz(t)}{dt} = Re\big(x^*(t)y(t)\big) - bz(t) \tag{3.5}$$

In these equations, $x(t)$ is the time evolution of the scaled electric field, $y(t)$ is the induced dipole moment (atomic polarization) and $z(t)$ is the population inversion. The parameter σ is the decay rate of the population inversion and of the photons in the laser cavity. The parameters r and δ are the pump rate and the detuning of the atomic resonance frequency from the optical frequency. The parameter b is the inversed decay rate of the induced dipole moment. In some configurations of the parameters leading to chaos, the three-dimensional representation of $x(t)$, $y(t)$ and $z(t)$ is composed of attractors along which the trajectories wrap [10]. An example with two attractors is shown in Fig. 3.2.

When the time constants of the relaxations are of the same order of magnitude, the three Lorenz–Haken differential equations must be taken into account. The lasers fulfilling these conditions are called Class C lasers [11]. For instance, gas lasers such as NH_3 lasers and Ne-Xe lasers are Class C lasers. Only Class C lasers can produce instabilities under free-running operation. Lorenz–Haken chaos in a free-running laser was achieved in an 81.5 μm NH_3 laser, in which low pressure and long wavelength combine to achieve chaotic conditions [12]. When there is a large discrepancy in the decay rates, it is possible to adiabatically eliminate some of the aforementioned equations while always keeping that for the electric field. When this elimination leads to two coupled equations for the dynamics of the electric field and that of the population inversion, this means that the laser under study is a Class B laser. Free-running Class B lasers are not prone to chaotic dynamics because they require one extra degree of freedom to exhibit deterministic chaos. However, they are easily destabilized by the introduction of external perturbations. If the equations for the field amplitude and the phase couple with each other through a perturbation, the laser

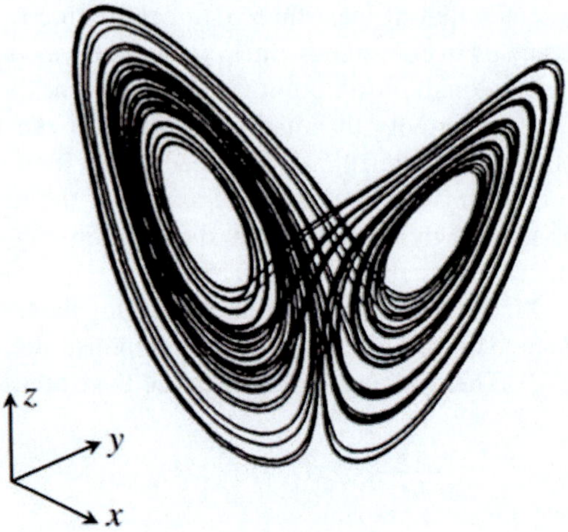

Fig. 3.2 Lorenz attractor when taking into account the Lorenz-Haken equations with parameter values of: $\sigma = 10$, $r = 28$ and $b = \frac{8}{3}$. The butterfly shape of this strange attractor may be one of the reasons Lorenz talked about this animal in order to illustrate the sensitivity to initial conditions when it comes to study chaotic meteorological phenomena [17]: "Does the flap of a butterfly's wings in Brazil set off a tornado in Texas?". Courtesy of [10]

must be described by the rate equations coupled with three variables. A laser coupled with three variables becomes a chaotic system and shows instabilities. Examples of external perturbations [13] are modulation of the laser parameters, external optical injection and various schemes of external optical or electro-optical feedback that will be described afterwards. One extra degree of freedom can also be achieved via Q-switching [14]. CO_2 lasers and laser diodes are examples of Class B lasers. Thorough experimental and numerical details of laser chaos have been reported in a CO_2 laser where loss modulation provides the additional degree of freedom to achieve chaotic trajectories [15]. Another typical feature of Class B lasers is a relaxation oscillation of the laser output that is observed for a step-time response when the population inversion does not follow the photon decay rate. This property explains why all the semiconductor lasers cannot be considered as Class B lasers. Indeed, some of them, such as QCLs, do not exhibit relaxation oscillations [16]. The last class of lasers requires two extra degrees of freedom in order to be destabilized and they are called Class A lasers. In other words, the differential equations for the polarization of matter and for the population inversion can be eliminated because the lifetime of photons in such laser medium is large enough compared with the other time constants of the relaxations. Examples of Class A lasers include, but are not limited to, visible He-Ne lasers and dye lasers. It is relevant to note that chaos is obviously not limited to optoelectronics or atmospheric phenomena [17]. Domains as varied as chemistry [18], predator-prey systems [19] or stock inventories [20] share the same peculiarities related to chaotic dynamics.

3.1.2 Chaotic Logistic Maps and Bifurcation Diagrams

In the case of a discrete system, we introduce a relationship with an event that becomes the new input of the system and causes the next events. This relationship is called discrete difference equations. We assume a non-linear response of the discrete system and write its output as $x(n)$ [8]:

$$x(n+1) = f\left(x(n); \mu\right) \tag{3.6}$$

In this equation, n is the step of the occurrence, $f(x)$ is the non-linear function, and μ is the system parameter. Generally, $x(n)$ is a vector and the parameter μ is a vector as well. Logistic mapping is the name of the mathematical relation for a discrete chaotic system. Using a quadratic function, we consider the logistic map, with a variable x and a parameter μ, given by:

$$x_{n+1} = \mu x_n (x_n - 1) \tag{3.7}$$

The important parameter in order to determine the behavior of x is μ. Stability and convergence to a fixed point are achieved for any initial value of x as long as the parameter μ is roughly less than 3. Oscillations can be retrieved when the parameter μ is larger than 3. With a further increase of μ, the system shows period-doubling oscillatory solutions and, finally, reaches irregular chaotic oscillations. Figure 3.3a (top chart) shows the chaotic output obtained when $\mu = 3.8$. These fluctuations are totally different from that observed in stochastic processes. The irregularity is simply derived from the deterministic difference equation in Eq. 3.7 and the irregularity is called chaos. One of the unique features of chaos is the sensitivity to initial conditions.

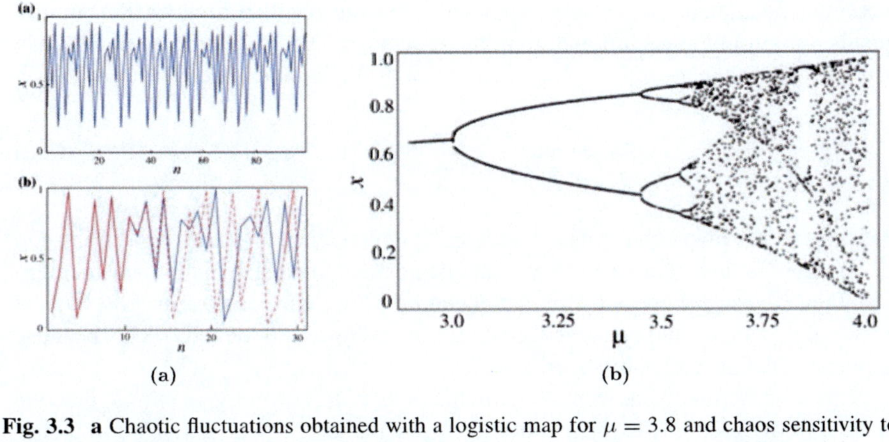

Fig. 3.3 **a** Chaotic fluctuations obtained with a logistic map for $\mu = 3.8$ and chaos sensitivity to the initial conditions after a few cycles; the blue curve corresponds to $x_0 = 0.12000$ and the red curve corresponds to $x_0 = 0.12001$. **b** Related bifurcation diagram gathering the extrema of the dynamics when varying parameter μ. Courtesy of [8]

A small discrepancy in the initial conditions leads to two different chaotic patterns after a few cycles, as shown in Fig. 3.3a (bottom chart) when $x_0 = 0.12000$ (blue) and $x_0 = 0.12001$ (red).

From the plot of the extrema of the variable x for the evolution of step n, we obtain a map for the parameter μ as shown in Fig. 3.3b. This map is defined as a bifurcation diagram. From this diagram, it is possible to see the evolution of the parameter x, starting with a stable behavior followed by periodic fluctuations for μ larger than 3, corresponding to the first bifurcation. Then, period-doubling oscillations occur for the increase of the parameter μ, and this corresponds to a new bifurcation in the diagram. Eventually, the output x displays completely irregular fluctuations at high values of the parameter μ, and this corresponds to the chaotic bubble in the diagram. This type of bifurcation is called a period-doubling bifurcation but is not the only route to chaos. Other routes for chaotic evolution exist depending on the configuration of systems and chaos parameters. For instance, quasi-period-doubling bifurcations and intermittent chaotic bifurcations [10] have already been exhibited. A bifurcation diagram is a key concept to understand how the dynamics of a system change with the parameters.

3.1.3 *Characterization of Chaos*

At first sight, a chaotic waveform can hardly be distinguished from a stochastic noisy timetrace and consequently, it is mandatory to have tools, such as the Lyapunov exponents (LEs), in order to characterize the observed dynamics [21]. The basic idea of the Lyapunov exponent λ is to measure the rate at which two originally nearby trajectories diverge in time, as we previously mentioned that two close trajectories are supposed to exponentially diverge in the case of chaos. As the direction of the maximum divergence or convergence locally changes on the attractor, the motion must be monitored at each point along the trajectory. The LE related to Eq. 3.6 can be defined from the absolute value of the slope of the logistic map $|f'(x_k)|$ as:

$$\lambda = \lim_{n\to\infty} \frac{1}{n} \sum_{0<k<n-1} ln|f'(x_k)| \tag{3.8}$$

The slope of the map determines the degree of dependence on initial conditions. On the one hand, at least one of the LEs must be positive to comply with the aforementioned sensitive dependence on the initial conditions for chaotic systems [22]. On the other hand, constant and periodic signals are linked to zero or negative LEs, because this means that nearby points converge.

We have described characteristics of chaotic dynamics through a simple discrete model, but these elements (bifurcation diagrams, LEs...) can be extended to a continuous media. The latter configuration is more appropriate to study the non-linear dynamics of lasers, as we saw that the Lorenz-Haken equations is not a set of discrete

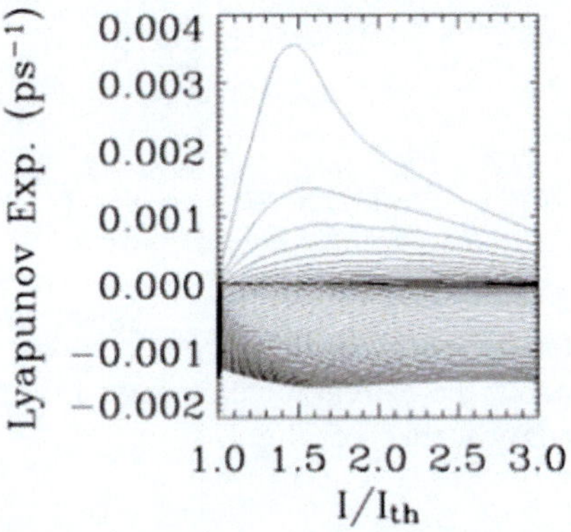

Fig. 3.4 Spectrum of the 50 largest Lyapunov exponents when varying the bias current of a semiconductor laser subject to optical feedback. Even if the largest Lyapunov exponent alone can give indications about the chaotic nature of a signal, a larger number of Lyapunov exponents are required to derive the Kolmogorov-Sinaï entropy and the Kaplan-Yorke dimension, which give further details about the chaos complexity of a signal. Courtesy of [21]

equations. Details about the transition from a discrete system to a continuous-time system can be found in Ref. [13]. Several methods for the calculation of the LEs from experimental time series have been developed [23–27]. In general, the number of LEs that can be retrieved corresponds to the number of system variables, and only the largest exponents are derived in order to determine if the system is chaotic, as shown in Fig. 3.4. However, a full computation of the Lyapunov spectrum (that is to say, all the LEs) allows a quantitative evaluation of the complexity of the dynamics. This evaluation is performed through the Kolmogorov-Sinaï entropy (KSE) [28] and the Kaplan-Yorke dimension (KYD) [29]. KSE is an estimation of the average loss of information rate. This means that KSE is inversely proportional to the time-interval over which the future evolution can be predicted. From the ordered Lyapunov spectrum, an upper bound of KSE is derived by taking the sum of all the LEs above zero:

$$KSE \leq \sum_{k|\lambda_k>0} \lambda_k \tag{3.9}$$

This formula is called Pesin inequality [30]. KSE either takes a finite positive value for chaotic dynamics or an infinite value for perfectly stochastic systems. A larger value of KSE implies more unpredictability for the system under study or, in other words, the larger the entropy, the larger the unpredictability of the system, which is a highly desired property to ensure security in chaos secure communications which will be developed hereafter. Another tool to evaluate the degree of chaos is KYD, which is a measure of complexity of chaotic attractors and which describes the amount of information required to locate the system in the phase space. If we assume that we have access to the ordered Lyapunov spectrum, and if we set j the number of degrees of freedom that satisfies the condition:

$$\sum_{1 \leq k \leq j+1} \lambda_k < 0 < \sum_{1 \leq k \leq j} \lambda_k \tag{3.10}$$

then KYD is defined by:

$$KYD = j + \frac{\sum_{1 \leq k \leq j} \lambda_k}{|\lambda_{j+1}|} \tag{3.11}$$

Other dynamical characterizations of chaotic fluctuations exist but will not be detailed here. We can cite for instance the correlation dimension [31], which is implemented through the Grassberger-Procaccia algorithm [32].

3.2 Chaos in Optoelectronics

3.2.1 Theoretical Model for External Optical Feedback

As we already briefly mentioned, several methods exist to trigger chaos in the output of a laser. In this section, we will focus on chaos obtained with external optical feedback (EOF), as illustrated in Fig. 3.5, because it was the scheme that was mainly studied in our experimental work about non-linear dynamics in QCLs. This section will be illustrated with recent findings in the dynamics of QCLs under EOF. All these phenomena were previously encountered in other semiconductor lasers with EOF, especially laser diodes, sometimes several decades ago.

The formalism of a semiconductor laser under optical feedback was originally proposed by Lang and Kobayashi [33]. In this case, only one round-trip in the external cavity is considered, which suggests a relatively small amount of feedback. In our experiments, we suggest that several round-trips can occur, as mentioned afterwards. If E is the slowly varying envelop of the complex electric field and N the carrier density of the upper laser state, the rate equations of the laser subject to optical feedback are expressed as:

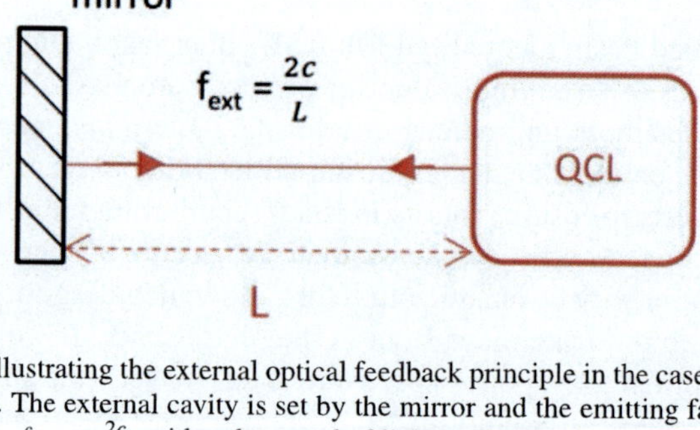

Fig. 3.5 Scheme illustrating the external optical feedback principle in the case of a semiconductor laser (here a QCL). The external cavity is set by the mirror and the emitting facet of QCL and the related frequency is $f_{ext} = \frac{2c}{L}$, with c the speed of light and L the external-cavity length

$$\frac{dN}{dt} = \frac{I}{q} - \frac{N}{\tau_c} - G(N)|E|^2 \tag{3.12}$$

$$\frac{dE}{dt} = \frac{1+i\alpha}{2}\left(G(N) - \frac{1}{\tau_p}\right)E + \frac{k}{\tau_{in}}E(t-\tau_{ext})e^{-i\omega_0\tau_{ext}} \tag{3.13}$$

where I is the bias current, q the electron charge, τ_c the carrier lifetime, τ_p the photon lifetime, $G(N)$ the gain per unit time, α the linewidth enhancement factor, ω_0 the free-running laser angular frequency and τ_{in} the internal-cavity round-trip time. The feedback appears in the last term of Eq. 3.13, with τ_{ext} the external-cavity round-trip time and k the feedback coefficient:

$$k = \frac{1}{\tau_{in}}2C_l\sqrt{f_{ext}} \tag{3.14}$$

where C_l is an external coupling coefficient that can be expressed for a FP-laser as:

$$C_l = \frac{1-R_2}{2\sqrt{R_2}} \tag{3.15}$$

with R_2 the reflection coefficient of the laser front facet subjected to the reinjection. In the case of a DFB-laser, the expression of C_l becomes much more complex and depends on the complex reflectivity at both laser facets [34]:

$$C_l = \frac{2(1-|\rho_l|^2)e^{-i\varphi_l}(q^2+\kappa^2)L^2}{i\kappa L(1+\rho_l^2) - 2\rho_l qL}\frac{1}{2qL - \sum\limits_{k=l,r}\dfrac{(1-\rho_k^2)\kappa L}{2iqL\rho_k + \kappa L(1+\rho_k^2)}} \tag{3.16}$$

In this equation, $\rho_k = |\rho_k|e^{i\varphi_k}$ with $k = l, r$, the complex reflectivity at the left or right facet, respectively (the right facet being the one towards the external cavity). L is the laser length, κ the coupling coefficient of the DFB grating and q can be expressed as $q = \alpha_{tot} + i\delta_0$ with α_{tot} the total internal losses and δ_0 the Bragg detuning, defined as the deviation between the lasing and the Bragg wavenumbers.

By writing $E = \sqrt{S}e^{i\phi}$, two rate equations can be deduced from Eq. 3.13 for the amplitude and phase of the electric field:

$$\frac{dS}{dt} = \left(G(N) - \frac{1}{\tau_p}\right)S + 2\frac{k}{\tau in}\sqrt{S(t-\tau_{ext})}\sqrt{S(t)}\,\cos(\omega_0\tau_{ext} + \phi(t) - \phi(t-\tau_{ext})) \tag{3.17}$$

$$\frac{d\phi}{dt} = \frac{\alpha}{2}\left(G(N) - \frac{1}{\tau_p}\right) - \frac{k}{\tau_{in}}\sqrt{\frac{S(t-\tau_{ext})}{S(t)}}\,\sin(\omega_0\tau_{ext} + \phi(t) - \phi(t-\tau_{ext})) \tag{3.18}$$

3.2.2 Consequences on the QCL's Macroscopic Parameters

Under steady state, the photon density S_s is constant $S_s = S_s(t) = S_s(t - \tau_{ext})$. Therefore, the term in the cosine, $\omega_0\tau_{ext} + \phi_s(t) - \phi_s(t - \tau_{ext})$, must be time insensitive and the steady-state phase ϕ_s can be written, without loss of generality, as $\phi_s = (\omega_s - \omega_0)t$.

Equation 3.18 under steady-state gives [35]:

$$\omega_s - \omega_0 = -\frac{k}{\tau_{in}} \left[\alpha \cos(\omega_s \tau_{ext}) + \sin(\omega_s \tau_{ext})\right] \tag{3.19}$$

This leads to several solutions for ω_s, those are the angular frequencies of the external-cavity modes, or fixed points. Among these frequencies, the laser will tend to operate on the mode with minimum linewidth, corresponding to the best phase stability [36].

The steady-state equation resulting from Eq. 3.17 gives:

$$G(N) - \frac{1}{\tau_p} + 2\frac{k}{\tau_{in}} \cos(\omega_s \tau_{ext}) = 0 \tag{3.20}$$

where ω_s is the solution of Eq. 3.19 that has the minimum linewidth.

Therefore, optical feedback has an influence on the threshold gain and hence on the threshold current of the laser, compared to free-running operation. The higher the feedback strength, the lower the threshold current of the laser, as can be seen in Fig. 3.6a.

As shown is Fig. 3.6b, some undulations appear in the light-current characteristic curve of a laser under EOF, as well as hysteresis. This phenomenon, explained in [37], is due to a competition between the internal-cavity modes (or to the laser frequency in

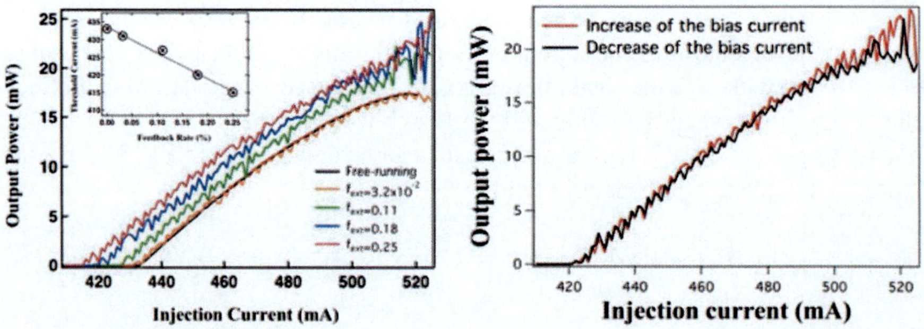

(a) (b)

Fig. 3.6 **a** L-I curves for several feedback rates at constant external-cavity length in the case of a QCL under EOF. The inset shows the corresponding threshold current as a function of the feedback rate (the dashed line is for visual guidance for the reader). **b** L-I curves at constant external-cavity length showing the hysteresis for a fixed feedback rate, when the bias current is first increased, then decreased. Courtesy of [54] the case of a DFB-laser) and the external-cavity modes. The heating of the active area further strengthens the hysteresis that appears between an increase and a decrease of the bias current.

3.2.3 Dimensionless Rate Equations with Optical Feedback

The dynamics of a laser diode subject to optical feedback is usually studied using dimensionless parameters [38], that can be extracted from the complex rate Eqs. 3.12 and 3.13.

In these equations, the above threshold gain per unit time can be expressed as:

$$G(N) = \frac{1}{\tau_p} + G_N(n_s - n_0) \tag{3.21}$$

where n_0 is the carrier density at threshold and G_N is the differential gain.

A normalization of the rate equations is performed by taking into account the photon lifetime τ_p and $s = t/\tau_p$ the normalized time constant. It is thus possible to define a new set of variables (Y,Z), with Y the normalized complex electric field and Z the normalized carrier density, by:

$$Y = \sqrt{\frac{\tau_c G_N}{2}} E \tag{3.22}$$

$$Z = \left(\frac{\tau_p G_N}{2}\right)(N - N_{th}) \tag{3.23}$$

The normalized rate equations can hence be written as:

$$\frac{dY}{ds} = (1 + i\alpha)ZY + \eta e^{-i\Omega\theta} Y(s - \theta) \tag{3.24}$$

$$T\frac{dZ}{ds} = P - Z - (1 + 2Z)|Y|^2 \tag{3.25}$$

with $\eta = k\tau_p$ the normalized feedback ratio, $\Omega = \omega_0\tau_p$ the normalized free-running frequency, $\theta = \tau_{ext}/\tau_p$ the normalized external-cavity round-trip time, $T = \tau_c/\tau_p$ the carrier-to-photon lifetime ratio and P the pump parameter, defined as:

$$P = \frac{\tau_p G_N N_{th}}{2}\left(\frac{J}{J_{th}} - 1\right) \tag{3.26}$$

In Eq. 3.26, the threshold current J_{th} and the carrier density at threshold N_{th} are linked by $J_{th} = N_{th}q/\tau_c$ with q the electron charge.

3.2.4 Influence of the α-factor

We used the Lang and Kobayashi model for a numerical analysis of the bifurcation diagrams when varying the value of the LEF. This model has initially been designed for diode lasers, but the conventional set of rate equations for QCLs can be approximated with good accuracy by the aforementioned model [39]. The parameters are set as described in Table 3.1 and are very close to the experimental parameters, so the simulations can be appropriately compared with the experimental results below. Figure 3.7 shows the bifurcation diagram for this set of parameters and a LEF of 0.5, 1.5 and 2.5, respectively. Bifurcation diagrams gather the intensity-output extrema with respect to the feedback strength, and allow analyzing the dynamics of the laser's output.

When the LEF equals 0.5 (Fig. 3.7a), the output of the laser remains constant even, for a feedback strength as high as 35%. Consequently, no dynamics are observed in this configuration which corresponds to values of LEF commonly found in QCLs [40, 41]. However, the lack of bifurcation is not compatible with the non-linear dynamics results we will unveil throughout this dissertation. Figure 3.7b shows the bifurcation diagram when the LEF is 1.5 and the other parameters remain constant. The output of the laser is constant until the feedback strength reaches 29%. Then, the Hopf bifurcation appears, meaning that the output enters an oscillatory state and eventually, the chaotic bubble is reached for a feedback strength of 30%. This bubble corresponds to the dense area in the diagram and means that the output exhibits several maxima and minima. This is related to a chaotic behavior. If the value of the LEF is increased to 2.5, the bifurcation diagram also displays a Hopf bifurcation, as shown in Fig. 3.7c, but for a lower feedback strength (19%). Then, the chaotic behavior is slightly different because not only one large chaotic bubble appears, like

Table 3.1 Physical and structural parameters used in the calculations

Parameter	Value
Laser length, l	2 mm
Reflectivity of the output facet, R_1	0.3
Reflectivity of the back facet, R_2	0.95
Internal cavity refractive index, n	3.2
Normalized cavity roundtrip time, θ	500
Coupling strength coefficient, C_l	0.64
Photon lifetime, τ_p	4.7 ps
Carrier lifetime, τ_c	4.7 ps
Carrier-to-photon lifetime, T	1
Differential gain, G_N	3.6×10^5 s^{-1}
Normalized pump parameter, P	4
Current pumping difference, $(J - J_{th})$	160 mA

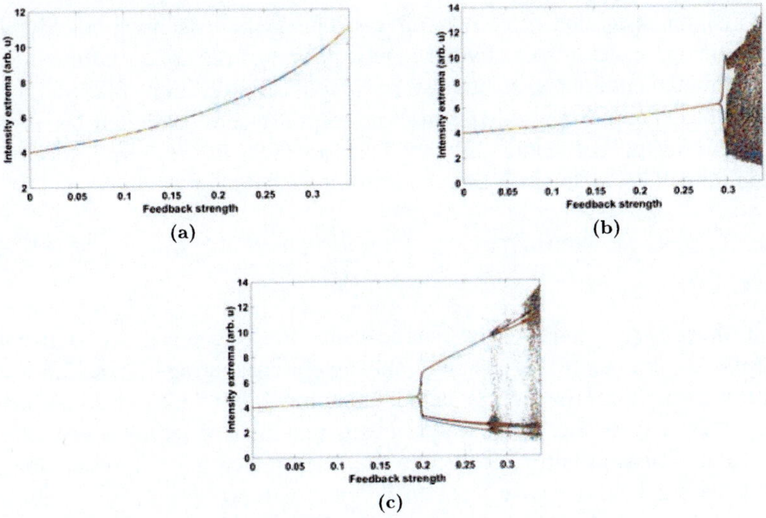

Fig. 3.7 Bifurcation diagrams for the parameters included in Table 3.1 and for various values of LEF. **a** $\alpha = 0.5$, **b** $\alpha = 1.5$ and **c** $\alpha = 2.5$

in the $\alpha = 1.5$ case, but several thin islands are retrieved. Each of them are separated by oscillatory states where the bifurcation diagram is composed of branches. The thin chaotic islands can for instance be found around a feedback strength of 20%, 28% and 33%. A comparison between the simulations and the results shown in the next subsections tends to confirm a LEF value high above zero since experimentally, the destabilization occurs for feedback ratios in the order of 1%. This is counterintuitive in a sense that the reciprocal space curvature of the subbands in a QCL should give a LEF very close to zero [42].

3.2.5 *Dynamical Regimes*

Following the discoveries of Lang and Kobayashi, several dynamics have been observed in semiconductor lasers under EOF: instability, bistability, self-pulsations, and coherence collapse states [43, 44]. In semiconductor lasers, EOF effects are frequently used for the control of oscillation frequency, selection of mode, and suppression of side-modes. For instance, it is possible to tune and stabilize the linewidth of lasers by applying a strong optical feedback [45–47]. Chirping of oscillation frequency can also be compensated by optical feedback. However, EOF can also have some downside effects, especially when the semiconductor laser shows unstable oscillations for a certain range of optical feedback levels. These effects may disturb a transmission in the case where the external cavity is defined by the emitting facet of the laser diode and the input of the fiber. In this case, an isolation of 40 dB is usually required to avoid reflection-sensitivity effects. This is the reason why the dynamics of semiconductor lasers induced by EOF are very interesting not only from the point of view of fundamental physics, but also for practical applications [48].

Cleaved facets are often used as a laser resonator in semiconductor lasers. Since the light is reflected perpendicularly to the laser facet in the cavity, the internal amplitude reflectivity r_0 is given by:

$$r_0 = \frac{\eta - 1}{\eta + 1} \tag{3.27}$$

where η is the refractive index of the laser material. For instance, the refractive index η of an AlGaAs semiconductor laser without any optical coating is about 3.6, and the amplitude reflectivity of the facet is calculated to be $r_0 = 0.565$. The corresponding intensity reflectivity is thus 32%, which means that 32% of the light generated by the stimulated emission is reinjected into the laser cavity, and the remaining light dissipates from the laser cavity [49]. In order to achieve high output power, the laser facets are coated appropriately by dielectric films. Then, the rear facet of the cavity usually has a high reflectivity of more than 90% and the front facet has a low reflectivity of less than 10%. This is quite different from other lasers where both facets have high reflectivities close to 100%. Despite such a dissipative laser structure, laser oscillations are still possible in semiconductor lasers because of the high efficiency of the conversion from pump to light.

There are many parameters to characterize instabilities and chaos in semiconductor lasers, such as the linewidth enhancement factor, the external-cavity length, the injection current above threshold, etc. Reference [50] extensively describes, with numerical simulations, the evolution of the bifurcation diagram of a QCL under EOF when varying these parameters. One of the most important and most useful parameters to figure out the characteristics is the feedback strength. The first investigation of semiconductor lasers with optical feedback was performed by Tkach and Chraplyvy [51], as can be seen in Fig. 3.8a. The instabilities of semiconductor lasers with optical feedback are categorized into the following five regimes, depending on the feedback fraction (which is not straightforward to evaluate experimentally):

- Regime I. Very small feedback and small effects. The feedback strength of electric-field amplitude is less than 0.01%. The linewidth of the laser oscillation becomes broad or narrow, depending on the feedback fraction [45].
- Regime II. Small, but not negligible effects. The feedback strength is less than 0.1%. Generation of the external modes gives rise to mode-hopping among internal and external modes [52].
- Regime III. This is a quite narrow region when the feedback strength is around 0.1%. The mode-hopping noise is suppressed and the laser restabilizes on the mode with the narrowest linewidth [51].
- Regime IV. This regime corresponds to moderate feedback strength around 1%. The relaxation oscillation becomes undamped and the laser linewidth is broadened greatly. The laser shows chaotic behavior and sometimes evolves into unstable

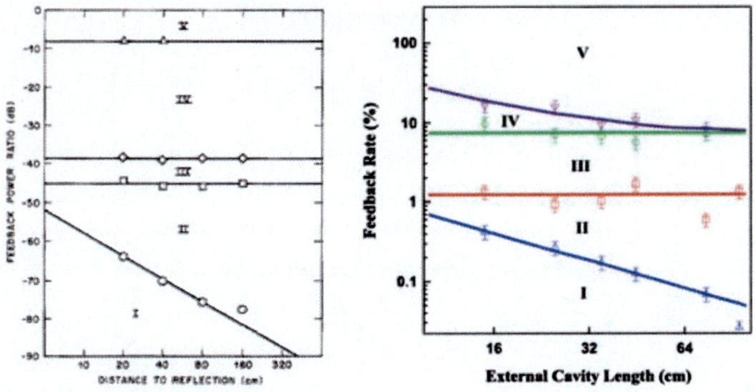

(a) (b)

Fig. 3.8 **a** First cartography of regimes of optical feedback as a function of the external-cavity length and the feedback ratio in a 1.55 μm distributed feedback QW-laser. **b** Cartography of the same regimes but in the case of a 5.6 μm QCL under EOF. Both semiconductor lasers show similar characteristics under EOF, even if their material structure is different. Courtesy of [51, 54]

oscillations in a coherence collapse state. The fluctuation level is enhanced greatly under this condition [44].

- Regime V. This is the strong feedback regime where the feedback strength is higher than 10% and that can be achieved only if the emitting facet of the laser is adequately coated. The internal and external cavities behave like a single cavity and the laser oscillates in a single mode. The linewidth of the laser is narrowed greatly [53].

This cartography remains valid for all single-mode semiconductor lasers so far, for which the influence of optical feedback has been studied, as can be seen in Figs. 3.8b and 3.9 for the case of QCLs. The feedback ratios at which the transitions between regimes occur remain of the same order of magnitude in terms of feedback strength [54], even if the third regime can sometimes become very narrow and difficult to locate, especially in the case of quantum dot lasers [55]. Over the years, this cartography has been complemented, especially to describe sub-regimes of the coherence collapse [56] or to address the small-cavity regime [57], defined when $f_r \tau_{ext} < 1$, with f_r the relaxation oscillation frequency and τ_{ext} the external-cavity round-trip time.

3.2.6 Coherence Collapse and Chaos

The term "coherence collapse" was introduced by Lenstra and co-workers in 1985 [44]. By considering the inter-dependence of amplitude and phase, a strong linewidth broadening as well as a huge decrease of the coherence length compared to the free-running case were obtained numerically, when the laser under study entered the

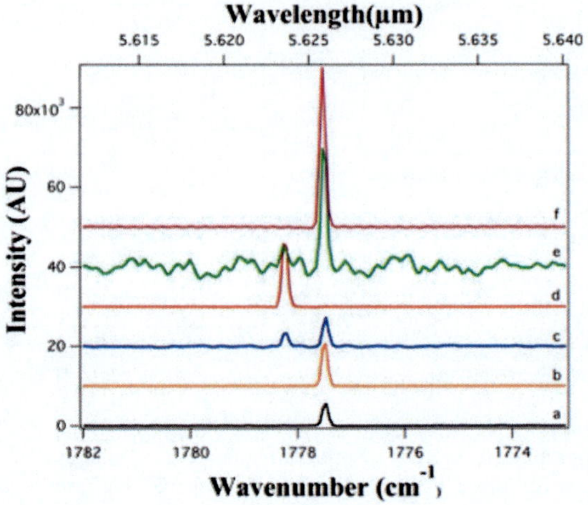

Fig. 3.9 Optical spectrum signatures of the five regimes in the case of a QCL under EOF. **a** Free-running case. **b** Regime I, $f_{ext} = 9.1 \times 10^{-4}$. **c** Regime II, $f_{ext} = 5.1 \times 10^{-3}$. **d** Regime III, $f_{ext} = 3.2 \times 10^{-2}$. **e** Regime IV, $f_{ext} = 0.13$. **f** Regime V, $f_{ext} = 0.25$. Courtesy of [54]

aforementioned regime IV. These findings were in good agreement with the experimental results. The first indication of the appearance of chaos within the coherence collapse regime originated from the analysis of the experimental electrical spectra of a laser, which was shown to match the power spectral density derived from the numerical time traces [58]. The two predicted routes to chaos, quasi-periodic or through period-doubling, were observed experimentally, the occurrence of one scenario or the other depending on the laser intrinsic characteristics and on the initial conditions.

3.2.7 Low Frequency Fluctuations

Low-frequency fluctuations (LFFs) is a type of chaotic fluctuations known as intermittent chaos and observed in lasers with optical feedback. LFFs were initially observed near the solitary laser threshold for high feedback ratios, but it was later proved that LFFs occur everywhere along the boundary between the optical feedback regimes IV and V discussed in the previous subsection. Indeed, due to the low internal reflectivity of the front facet of semiconductor lasers, LFFs are observed not only for low injection currents close to the laser threshold, but also for higher injection currents [59], as shown in Fig. 3.10. As a general trend, the frequency of LFFs linearly increases with the increase of the bias injection current [60], as can be seen in Fig. 3.11. The study of LFFs is very important from the viewpoint of practical applications, since LFFs induce more fluctuations in the laser output power. A typical feature of LFFs is a sudden power dropout with a following gradual power

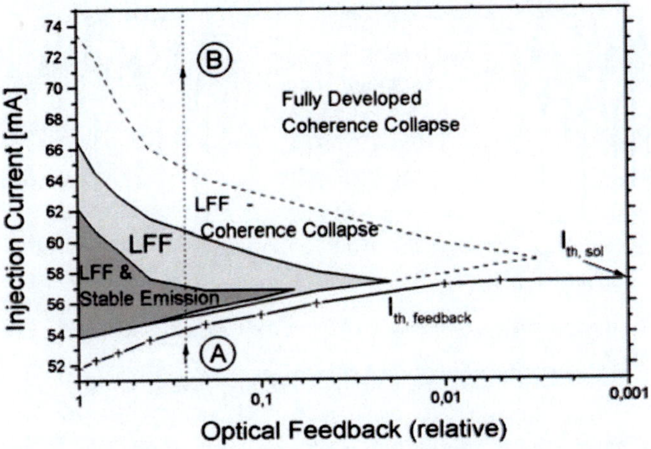

Fig. 3.10 Dynamical behavior of a semiconductor laser subject to optical feedback in feedback-current space showing that LFFs are not restrained to near-threshold cases. The LFF regime is depicted in light gray. The dark-gray region embedded in the LFF regime corresponds to the region of coexistence of the stable emission state and the LFF state. The white region encompassed by the dashed line corresponds to the continuous transition between the LFF regime and the fully developed coherence collapse regime. Courtesy of [59]

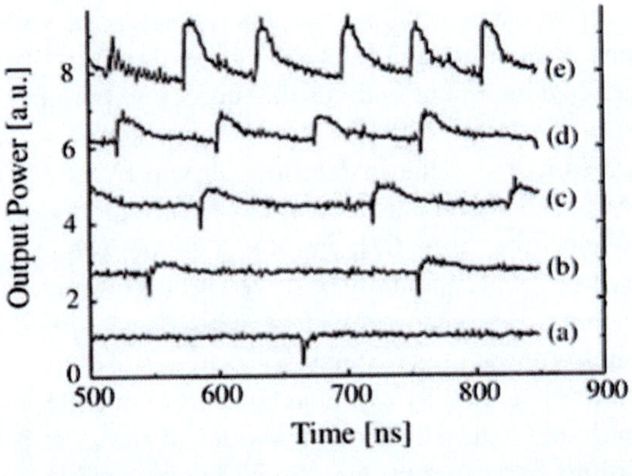

Fig. 3.11 Time series of LFFs showing the influence of bias currents on the frequency of the LFF pattern in the case of a semiconductor laser under external optical feedback. The laser is pumped at **a** $J = 1.20 \times J_{th}$, **b** $J = 1.34 \times J_{th}$, **c** $J = 1.48 \times J_{th}$, **d** $J = 1.66 \times J_{th}$, **e** $J = 2.08 \times J_{th}$, respectively. Courtesy of [60]

recovery, or sometimes reciprocally as illustrated in Fig. 3.12 in the case of a QCL under EOF. LFFs occur irregularly in time, depending on the system parameters, and the main frequency of LFFs is usually of the order of MHz to a few dozens of MHz. Since the main frequency of LFFs is much lower than that of ordinary chaotic fluctuations, these phenomena are called low-frequency fluctuations. Indeed, regular chaotic fluctuations are related to the relaxation oscillations of the laser (if any),

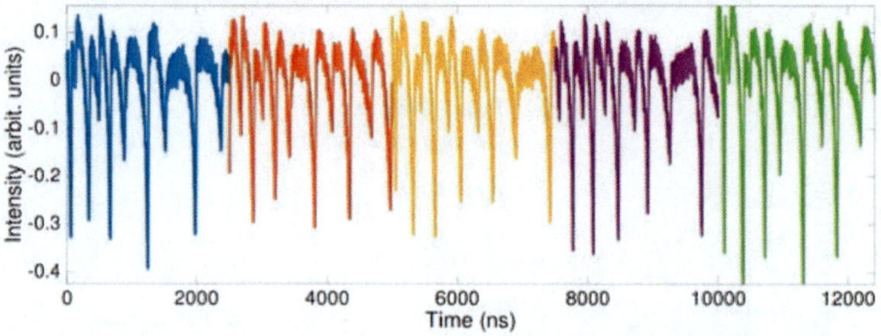

Fig. 3.12 First experimental observation of LFFs in the case of a QCL under external optical feedback. Courtesy of [74]

and the frequency of these oscillations is often at several GHz [61]. When temporal waveforms of LFFs are observed by a fast digital oscilloscope, they seem to be continuous signals. However, it is proved that LFFs bind very fast time structures within the waveform, consisting of series of fast pulses in the order of subnanoseconds. Indeed, this fast pulsation has been experimentally observed by using a streak camera [62]. To summarize, LFFs, which are first recognized as fluctuations of the laser output power with a low frequency, are made of three components of different time-scales. One is a component of a low-frequency fluctuation with a period around MHz, the second one is a component related to the external-cavity length, the third one is a high frequency component with a subnanosecond period.

Various models for the origin of LFFs have been proposed since the 80ies [63]. One of them describes LFFs as the instabilities driven by noise in the non-linear laser system [64–66]. Other ones are explained by competitions between modes in a multimode model of the laser [67], and the crisis between two attractors [68]. However, LFFs have been experimentally observed in single-mode lasers with optical feedback, and they have also been proved to appear in a single-mode model by numerical simulations without noise effects.

Phenomena of LFFs are generally explained by employing the model that the laser output power hops around external- and anti-modes of the laser oscillations due to saddle node instabilities generated by the optical feedback [69]. We derive a set of two equations in the case of external- and anti-modes when taking into account the phase space of the oscillation frequency and the carrier density. In semiconductor lasers with optical feedback, we obtain the steady-state solutions from Eqs. 3.19, 3.20 to 3.21 as:

$$\Delta\omega\tau_{ext} = -\frac{\kappa\tau_{ext}}{\tau_{in}}\sqrt{1+\alpha^2}sin\left(\omega_s\tau_{ext} + tan^{-1}(\alpha)\right) \tag{3.28}$$

$$\Delta n = -\frac{2\kappa}{\tau_{in}G_n}cos\left(\omega_s\tau_{ext}\right) \tag{3.29}$$

where $\Delta\omega = \omega_s - \omega_0$ and $\Delta n = n_s - n_0$. Figure 3.13a shows the ellipsoid of the distribution of the modes in the phase space, which were calculated from the two previous equations. This corresponds to the case of a long external-cavity length with a high level of external feedback. Therefore, the modes are densely distributed. For the case of short external-cavity length with a low external feedback, the mode distribution would be sparse. The black dot at the center of the ellipsoid is the mode of the solitary oscillation. During LFF, the build-up process is based on a mode-switching towards the mode with maximum gain, located at the lower extremity of the ellipsoid. However, a so-called crisis [69] occurs during the process, corresponding to a collision between a quasi-attractor, that is to say an external-cavity mode that temporarily looses its stability due to the advent of chaos, and the associated anti-mode. Because of this crisis, the Δn value will increase, and in order to restabilize, the laser frequency will shift until it meets the ellipsoid again on an anti-mode, and the stabilization then occurs on the corresponding low-power external-mode of the cavity.

As already mentioned, LFFs have three time-scales of fluctuations. The first one is a fast pulse-like oscillation of the order of subnanoseconds. When the laser is switching close to one of the external-modes, the carrier density fluctuates with large amplitude and it reaches the associated anti-mode, while the phase stays almost constant. This corresponds to a series of fast pulsations. The time-scale depends on the fluctuations of the carrier number in the active layer. The second one is the transition time between the successive external modes in the stepwise time recovery process. This time-scale is related to the external-cavity length and, for example, it is of the order of nanoseconds for an external-cavity length of several dozens of centimeters. The third time-scale is the time constant for power dropouts. This time-scale is usually of the order of microseconds, but the period is not fixed, as has already been discussed. The whole process, with its relation to the LFF waveform, is detailed in Fig. 3.13b. The authors assumed the steady-state condition to calculate the external- and anti-modes in Fig. 3.13a. However, the occurrence of LFFs is of dynamic nature, so that the discussion developed here may not be applicable for such dynamic states in the strict sense. Nonetheless, the approach can explain the overall features of the LFF dynamics, because chaos itself is generated from small perturbations to the initial state of the laser oscillations. This explanation allows understanding the origin of LFFs, and also concluding that the presence of LFFs in the dynamics of a laser under EOF is a manifestation of deterministic chaos. Many different phenomena can lead to slow fluctuations in a laser diode under optical feedback, such as a misaligned optical cavity [70]. In order to distinguish LFFs amid these fluctuations, it is possible to derive the largest Lyapunov exponents and see if they have a positive value, as already mentioned. Another method exists to characterize this kind of deterministic chaos and it is based on the study of the statistics of the period between two consecutive drop-outs [71–74]. As shown in Fig. 3.14 in the case of a QCL under EOF, the statistics of the LFF is characterized by a death area, standing for a minimum period of typically $10 \times \tau_{ext}$ below which no events occur. Moreover, it has the shape of a decaying exponential at low injection current, and a second

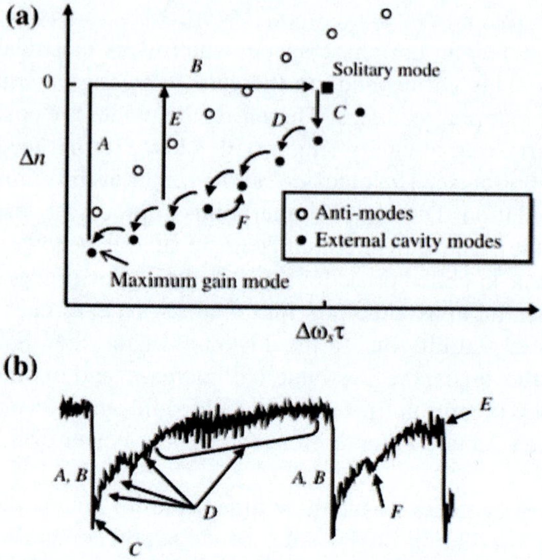

Fig. 3.13 **a** External- and anti-modes in the phase space of frequency and carrier density according to Sano's model [69]. Only the lower half of the ellipsoid of the mode distribution is shown in the graph. **b** Timetrace corresponding to the whole LFF process presented in (**a**). Courtesy of [8]

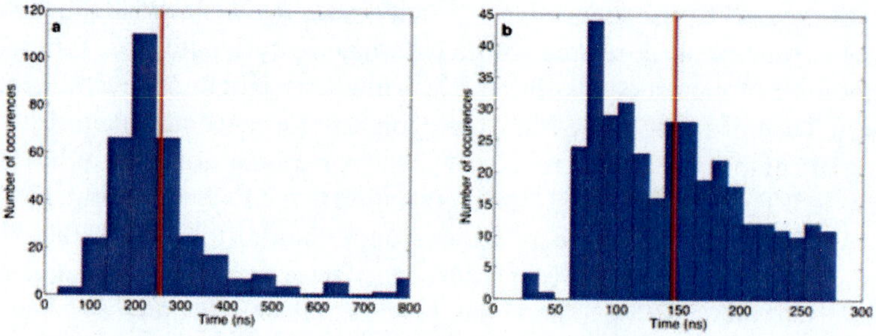

Fig. 3.14 Statistics of the time-interval between LFF dropouts in the case of a QCL under EOF. **a** corresponds to a feedback strength of 2.7% while **b** corresponds to a feedback strength of 11%. The solid vertical line indicates the average value. Courtesy of [74]

maximum pops up for stronger injection currents. Such statistical distribution is one of the characteristics of LFFs and allows deducing the presence of chaotic behavior in a laser under EOF.

3.2.8 *Applications of Semiconductor Laser Chaos*

In the following chapter, the advantages of chaos for secure communications will be thoroughly studied. However, other applications exist and will be briefly described in this subsection, as they could be promising in future broadband technologies such as autonomous vehicles [75]. Chaos can indeed be used in order to improve the efficiency of LIDAR (light detection and ranging). LIDAR is similar to RADAR (radio detection and ranging) in the detection process but offers increased resolution and better field of view due to shorter wavelengths. This is of paramount importance in order to ensure a reliable detection in an autonomous car driving on a road with a large incline, for example. Another advantage of LIDARs, compared to video cameras, is that LIDAR gives access to parameters of first importance such as the distance of the detected targets, as well as their speed thanks to Doppler effect. There are two categories of LIDARs: one based on lasers with ultra-short pulses and the second one based on frequency chirp. For the latter, the information about the target are recovered from the correlation between the back-reflected signal on the target and the delayed emitted signal. Several methods can be implemented in order to generate a signal with no repetition of the delayed outcome, because a repetition could lead to the measurement of a wrong target's distance. Lin and Liu developed chaotic LIDARs [76] and chaotic RADARs [77], taking advantage of the chaos properties of laser diodes in order to create a signal at high frequency, which means high resolution, and with no dependence on any previous outcome. The detection step can be of the order of the centimeter and the range can be extended to several meters, as visualized in Fig. 3.15. Furthermore, the use of chaotic signals instead of modulated waves leads to detection systems resistant to noise and jamming [78].

Another application strongly supported by optical chaos is physical random number generation (RNG) [79]. Random numbers are vital ingredients in many applications ranging, in a non-exhaustive list, from computational methods such as Monte Carlo simulations and programming [80], to the large field of cryptography for the generation of enciphered messages, as far as to commercial applications like lottery games and one-arm bandits [81]. Even though different by nature, chaos and randomness share a common feature in that they produce entropy, as we saw it in the case of chaos with KSE. Chaos in semiconductor lasers thus provides an ideal physical source of random bits, as it combines outcome-unpredictability with no dependence on any previous outcome. These two requirements are of paramount importance for RNG [82]. Also, chaotic semiconductor lasers may produce a large number of positive LEs, whose magnitudes relate to the bandwidth of the retrieved chaotic frequencies [83]. Using chaos from a laser diode, random bits can therefore be produced at very high bit-rates [84], exceeding those obtained with other physical sources of entropy, including quantum random number generators, which are currently limited at approximately one Gbits/s [85] and may achieve a few dozens of Gbits/s in the coming years [86, 87]. After the first successful demonstration in 2008 [88], the field of RNG using chaotic laser diodes has benefited from several developments, in terms of sampling rates and number of generated bits, as shown in Fig. 3.16. Recent experiments have confirmed the possibility to excess a sampling rate of one Tbits/s [89, 90].

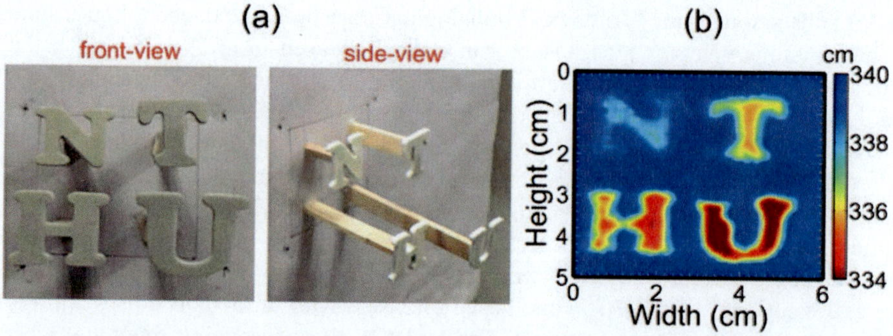

Fig. 3.15 **a** Picture of several letters placed at more than 3 meters from the detection system (with semiconductor laser and detector) and **b** its corresponding 3D image obtained by the pulsed-chaos LIDAR. The colorbar shows the relative range from the LIDAR. Courtesy of [78]

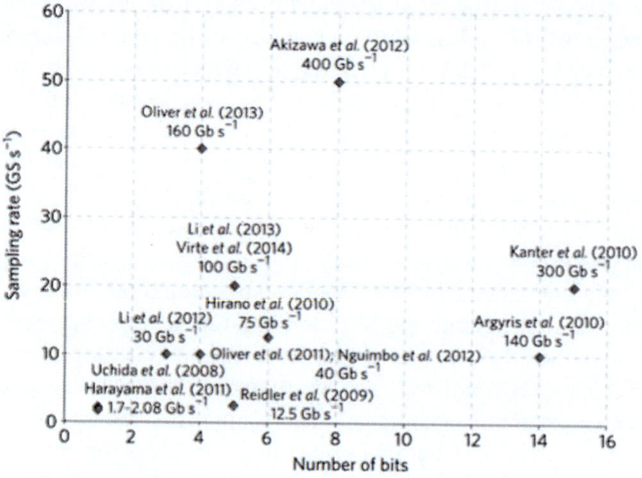

Fig. 3.16 Evolution of the performances of physical random number generation with semiconductor laser chaos, between 2008 and 2014. Courtesy of [79]

3.3 Chaos in QCLs

After the discoveries related to deterministic chaos for several semiconductor lasers in the 80ies and the 90ies, it was relevant to extend these notions to QCLs because their emission wavelength in the mid-infrared make them useful for several applications which cannot be addressed with near-infrared and visible-light semiconductor lasers. However, QCLs do not exhibit relaxation oscillations and they are renown for their low α-factor compared to other semiconductor lasers which, overall, brings stability to QCLs subject to external perturbations. It was thus not straightforward to experimentally retrieve chaotic dynamics with such configuration [91].

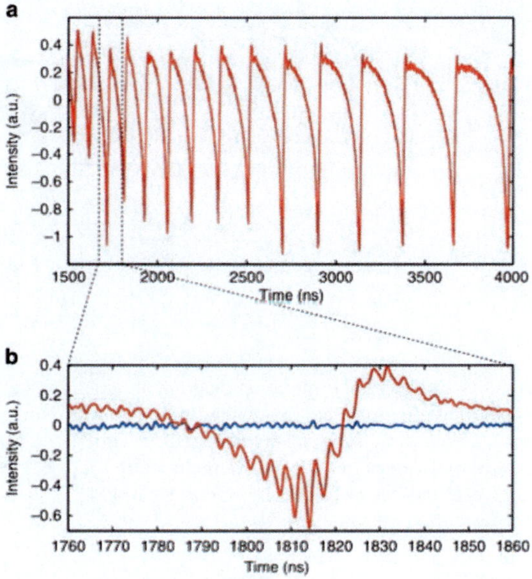

Fig. 3.17 **a** Pioneering experimental results regarding chaos in QCLs under EOF. **b** Focus on one of the LFF dropouts revealing the contribution of the external-cavity frequency. Courtesy of [74]

The first experimental proof of LFFs in QCLs was given by Jumpertz and co-workers in 2016 [74]. At both near-threshold pump current and high feedback strength, they were able to follow the evolution of the non-linear dynamics, with the advent of oscillations at the frequency of the external cavity and then, at higher feedback strength, LFFs containing both the frequency of the external cavity and the dropout signature, as can be seen in Fig. 3.17. Not only they exhibited the temporal chaotic waveforms, but also they performed thorough numerical simulations based on the Lang and Kobayashi equations already mentioned. When taking into account the peculiarities of QCLs, namely a low α-factor and a very low carrier-to-photon lifetime, the simulations showed a Hopf bifurcation followed by a chaotic bubble, and were compatible with the experimental results, as can be seen in Fig. 3.18. Although of prime importance because it was the first proof-of-concept of chaos in mid-infrared QCLs, these preliminary results left many questions unanswered. The non-linear dynamics were observed only close to the current threshold for a quasi-continuous bias and at room temperature. In order to have a comprehensive overview of the phenomenon and to take advantage of it for applications, it was necessary to determine if chaos was more developed at high or low temperature, with a continuous bias or a quasi-continuous bias and for a bias current close to threshold or far from threshold.

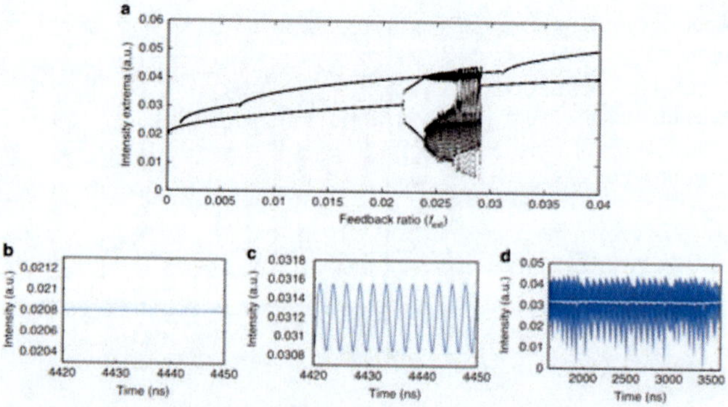

Fig. 3.18 Numerical bifurcation diagram obtained for a QCL under EOF when using the Lang and Kobayashi equations, and associated time series. **a** Numerical bifurcation diagram. **b** Time trace for $f_{ext} = 0.11\%$, showing stable signal. **c** Time trace for $f_{ext} = 2.14\%$, showing oscillations at the external-cavity frequency. **d** Time trace for $f_{ext} = 2.59\%$, showing both LFF and oscillations at the external-cavity frequency. Courtesy of [74]

3.3.1 Experimental Apparatus

As previously mentioned in order to illustrate EOF phenomenon, Fig. 3.19 shows the experimental setup we worked with in order to provide the experimental results contained in this chapter and the next two. The experimental apparatus is made of two branches. On the one hand, there is a feedback path with a mid-infrared polarizer and a gold-plated mirror placed on an accurately-moving cart. This mirror, together with the emitting facet of the QCL, defines the external-cavity length which is one of the main parameters of EOF. The polarizer is the cornerstone device for tuning the amount of optical feedback knowing that the QCL wave is TM polarized: the angle of the polarizer defines the feedback strength. On the other hand, in the detection path, we use a high bandwidth mid-infrared detector (Vigo PEM Mercury-Cadmium-Telluride; MCT) operating at room temperature. The signal retrieved from

Fig. 3.19 Experimental setup used to study a QCL under EOF. NPBS: non-polarizing beam-splitter, MCT: Mercury-Cadmium-Telluride, OSCI: digital oscilloscope, RSA: electric spectrum analyzer

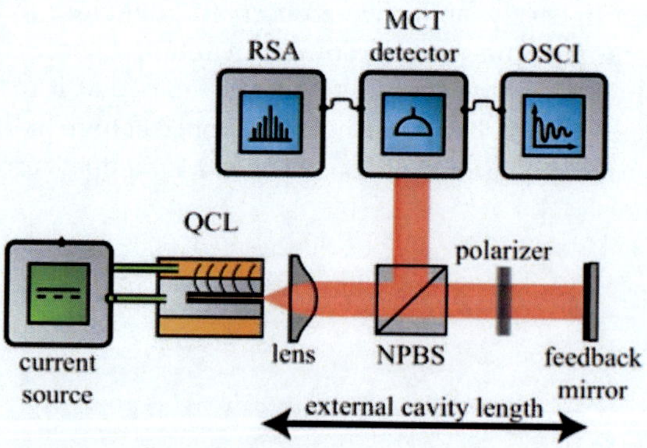

the MCT detector is amplified with a wide-bandwidth low-noise amplifier, in order to overcome the background noise. The signal is then subsequently analyzed with a real-time digital oscilloscope and an electric spectrum analyzer. For an external-cavity length between 30-60 cm, the related external-cavity frequency is theoretically between 250-500 MHz. A 60/40 mid-infrared beam-splitter then splits the focused laser beam into both paths. Focusing is achieved with a lens in front of the laser. Two different setups are implemented, based on the measurement temperature. When the laser is studied at room temperature, the QCL package is horizontally clamped over an indium foil and a copper mount with a Peltier module for temperature control. In that configuration, the wave hitting the beam-splitter is P-polarized and the transmission of the beam-splitter at this wavelength is about 60%. Low-temperature measurements down to 77 K are implemented in a cryostat. The QCL is vertically clamped over a copper mount, the latter being placed inside a vacuum chamber in order to insulate from the outside. A heater and a temperature controller inside the vacuum chamber allow a better control of this key parameter. The cryostat has an output covered with a ZnSe window made for mid-infrared light, and a focusing lens is placed between this window and the beam-splitter. The wave hitting the beam-splitter is S-polarized and the maximum achievable transmission is 35% because the QCL is vertically inserted inside the vacuum chamber. In order to minimize the environmental perturbations such as acoustic and mechanical noises, the laser is mounted on a suspended optic table. Consequently, the feedback mirror remains immobile and the observed pattern cannot be related to self-mixing effects. Furthermore, the feedback ratios applied to the QCL are well above the ones required for self-mixing interferometry [92]. The current source in this experiment is a low-noise source and the continuous bias delivered by the source can be modulated with an external signal from a waveform generator. It is relevant to note that the low-noise source has a modulation bandwidth of 3 MHz. Hence, the measurements carried out above this value lead to a forcing with an amplitude that is smaller than that expected because of the low-pass filter embedded in the current source.

3.3.2 *Quasi-continuous Wave*

Our first efforts to fill the gaps in chaos mapping in QCLs drove us to the study of non-linear dynamics at high injection currents, because if mid-infrared chaos is to be used for long-range secure communications, then it is mandatory to drive the laser at a point where it is more likely to generate a high output power. Moreover, LFFs have already been observed far from threshold in other semiconductor lasers and this further motivated our studies.

Figure 3.20 shows the evolution of the waveforms with the feedback strength, as well as the associated bifurcation diagram for a quasi-continuous bias (pulses of 2 μs every 10 μs) at 290 K. The first μs of each time trace is removed because it corresponds to a transient state. When varying the amount of external optical feedback by means of the mid-infrared polarizer, several regimes such as stable operation

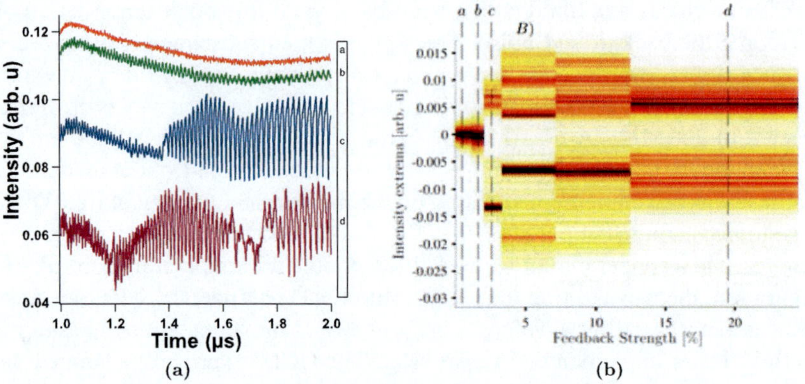

Fig. 3.20 Experimental time traces when the laser is pumped in the quasi-continuous mode **a** for several values of the feedback strength and related bifurcation diagram **b** when the temperature is 290 K. **a** 0.05%, **b** 1.1%, **c** 3.3%, **d** 24.6%; For low feedback strengths, the signal is nearly continuous and then, with increasing feedback strength, evolves towards non-linear patterns such as oscillations and LFFs

and LFFs appear. Studying these non-linear phenomena is essential for practical applications of such lasers into a communication system. However, it requires the optical path to be very stable, thus giving the advantage to room-temperature setups compared with cryostat setups, which are never really stable because of the working vacuum pump. It can be assured that feedback is at its maximum when the amplitude of the spikes, described in the next section, is also at its maximum. It is important to emphasize that the symmetry of the pattern needs to be carefully retrieved when varying the feedback strength. Indeed, restabilization can occur in semiconductor laser with EOF [74], meaning that a high feedback strength can lead to a case where the signal is constant and the RF spectrum is flat. We actually never observed such a behavior in the case of a quasi-continuous bias. Being able to precisely assess the golden mirror tilt giving the maximum feedback strength leads to phenomena which are highly reproducible.

In Fig. 3.20a, the lowest value of feedback strength displays a stable signal corresponding to free-running operation. Oscillations start appearing for feedback ratios above 1.1%, and the transition between oscillations and the LFF regime can be observed for a feedback ratio of 3.3%. The deterministic pattern becomes then more complex until the maximum feedback ratio of 24.6% is obtained. The related bifurcation diagram is plotted in Fig. 3.20b and displays the extrema of the experimental time traces for a selection of feedback strengths. Increasing the feedback ratio leads to changes in the dynamics observed in the time traces. The dashed lines correspond to the feedback strengths at the far-right of the solid lines, but are placed in the middle of the solid lines for visual guidance for the reader and represent the analysis of the time traces of Fig. 3.20a.

The bifurcation diagram plotted in Fig. 3.20b shows a Hopf bifurcation for a feedback strength of 3.3%. The corresponding time-periodic oscillations for higher

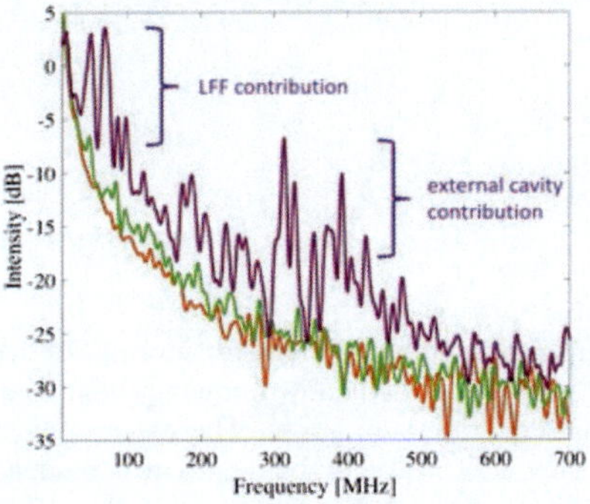

(a)

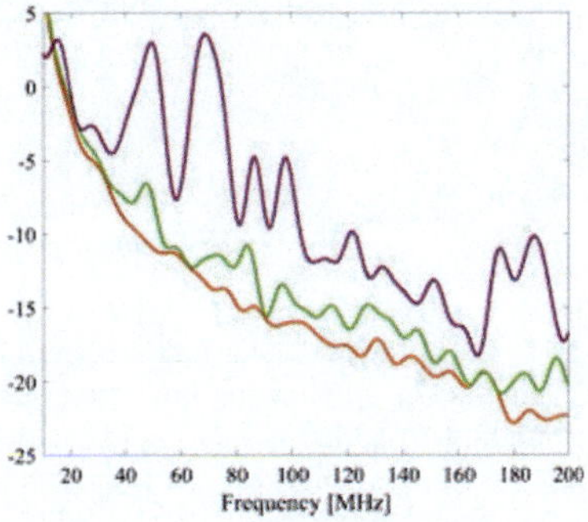

(b)

Fig. 3.21 RF spectra related to Fig. 3.20a, b, d and obtained with a fast Fourier transform (colors remain the same). **a** Analysis of the whole bandwidth. Only the purple curve (corresponding to a high feedback strength) shows the contribution of the external-cavity frequency around 330 MHz. A small discrepancy between the theoretical cavity frequency and the experimental one has already been pointed out [74]. **b** Close-up on low frequencies, the green curve (low feedback strength) is characterized by small bumps around 42 and 86 MHz, contrary to the orange curve (zero feedback strength). This supports the claim that the frequency of the pattern seen in Fig. 3.20b may be related to multiple reflections in the external cavity

feedback ratios have a frequency of nearly 50 MHz, which is ten times lower than the theoretical frequency of the external cavity ($f_{ext} \approx 430$ MHz for an external cavity of 35 cm) when oscillations are expected to occur [74]. This behavior may be due to multiple reflections since the laser is operated at $J = 2.2 \times J_{th}$, which corresponds to the current leading to the maximum output power. Figure 3.21 shows that the external-cavity frequency is missing for a configuration close to the Hopf bifurcation. Yet, this figure demonstrates that the frequency of the external cavity can be found in the LFF pattern for high feedback ratios. When increasing the feedback strength above 3.3%, the intensity extrema spread over a wide range of values corresponding to the onset of a chaotic pulsing output combined with fast pulses, which is represented by solid lines around the zero intensity value in the bifurcation diagram. In that experimental configuration, the transition between the oscillations and the LFF regime is characterized by the coexistence of the two regimes as shown in Fig. 3.20a, plot c). Another relevant conclusion that can be drawn is that the frequency of the LFF dropouts is higher in the case where the laser is biased at 2.2 times the threshold current compared to the case described in Ref. [74] and corresponding to an injection current close to threshold. Indeed, this frequency is comprised between 1 and 10 MHz according to Fig. 3.17 whereas this frequency is in the order of 30-50 MHz in Fig. 3.20a, plot d). That is in good agreement with similar findings in other semiconductor lasers, as already described in Fig. 3.11. In our experiment, the frequency of the external cavity is 430 MHz but the frequency of the fast oscillations is 330 MHz, as can be seen in Fig. 3.21a for the purple curve. This discrepancy between the two values has already been observed previously [50] and in both cases, the retrieved frequency is always lower than the expected frequency, as shown in

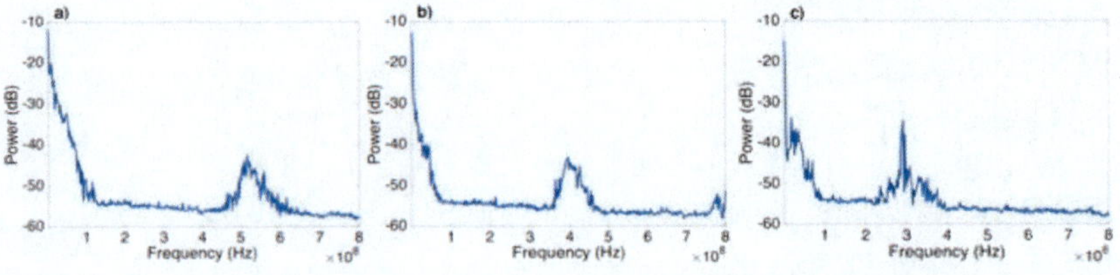

Fig. 3.22 Experimental electrical spectrum for a feedback strength of 3% and for several external-cavity lengths (L_{ext}), showing both the external-cavity frequency and a low frequency component corresponding to LFFs. **a** $L_{ext} = 25$ cm. **b** $L_{ext} = 35$ cm. **c** $L_{ext} = 45$ cm. The external-cavity frequency is always a bit lower than expected. For instance, $L_{ext} = 25$ cm should lead to a frequency of 600 MHz while the experimental result displays a peak at 520 MHz. Courtesy of [50]

Fig. 3.22. Eventually, we also performed the first analysis of LEs in the case of a QCL in order to assess the chaotic nature of the dropouts we observed. Figure 3.23 shows the phase portrait of the time trace corresponding to the highest level of feedback (Fig. 3.20a, plot d)), with and without noise filtering. The phase portrait of the temporal waveform shows a rich attractor structure. This figure also depicts the four largest LEs and it is possible to see that two LEs are positive, thus ensuring that the fluctuations are chaotic. This confirms the claims of the first observation of chaos in QCLs under EOF, where the chaotic nature was tested with a statistical analysis of the time-intervals between the dropouts in place of a LEs analysis.

In order to get a global picture of the chaos phenomenon in a QCL with external optical feedback, we also studied the influence of temperature in the case of a quasi-continuous bias. We decided to focus on the minimum feedback strength leading to LFFs so that we can see if a low temperature is more prone to chaos dynamics. Figure 3.24 shows our results (in light blue) for three temperatures. The trend is that the lower the temperature, the lower the feedback strength giving rise to LFFs, so it would be relevant to operate the laser at cryogenic temperature in order to improve applications relying on chaos with QCLs. However, the bulkiness of cryogenic systems may not be compatible with versatile applications, and the vibrations of the cryostat will always remain an issue, as already noticed. The best option seems to be a trade-off between cryogenic temperature (better for chaos) and room temperature (better for the stability of the beam) and therefore, Peltier cooling will be used in the following experiments as often as possible. Such a difference in the chaotic behavior between cryogenic temperature and room temperature can be attributed to the increase of the carrier-to-photon lifetime ratio. At room temperature, the latter is typically in the order of 0.1 in a QCL whereas it is four orders of magnitude larger in interband lasers [93]. As we already saw in 2.1.3, the carrier lifetime can be approximated via the upper-state lifetime τ_c, defined as follows:

$$\frac{1}{\tau_c} \approx \frac{1}{\tau_{31}} + \frac{1}{\tau_{32}} \tag{3.30}$$

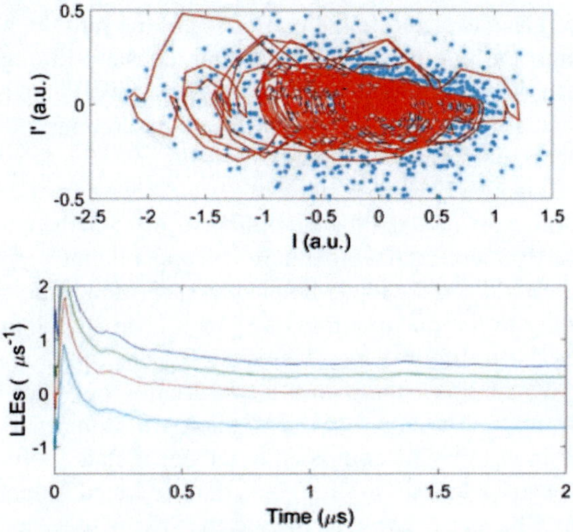

Fig. 3.23 System dynamics analysis through LEs for the maximum feedback strength at 290 K. The upper diagram is the phase portrait and the lower diagram is the Lyapunov spectrum for the four largest LEs calculated. As the time evolves, curves in the lower panel converge to the values of such exponents. For the calculated spectrum, the curves converge to values $\lambda_1 = 0.485\,\mu s^{-1}$, $\lambda_2 = 0.284\,\mu s^{-1}$, $\lambda_3 = -0.021\,\mu s^{-1}$ and $\lambda_4 = -0.656\,\mu s^{-1}$. Blue dots on the upper diagram are the retrieved values of the derivative of the laser's intensity I' as a function of the laser's intensity I, and the red curve represents the phase diagram after noise filtering

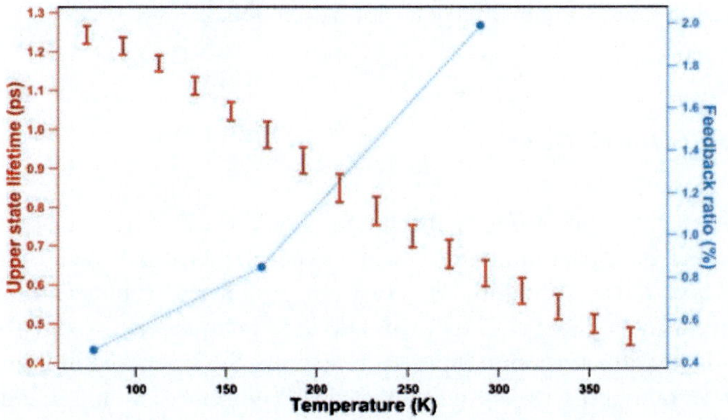

Fig. 3.24 Simulation of the upper state lifetime evolution with temperature for the QCL under study (red) as well as the required critical feedback (light blue) leading to LFF emergence for three temperatures: 290, 170 and 77 K; the dashed blue line is for the visual guidance of the reader

with τ_{32} the time constant related to the carrier scattering into the lower laser level and τ_{31} the one into the bottom level with a time constant through longitudinal-optical phonon emissions. The upper-state lifetime is investigated with a custom heterostructure simulation software, named METIS, based on semi-classical Boltzmann equations with thermalized subbands [94]. Figure 3.24 shows the evolution of τ_c from 373 K down to 73 K with steps of 20 K. Other numerical studies using a density matrix transport model showed similar results for the carrier lifetime of mid-infrared QCLs [95]. For each temperature, the upper bound represents the value of τ_c just above threshold and the lower bound represents the value of τ_c close to the current value leading to the maximum output power. Numerical simulations show that the mean upper state lifetime increases from 0.47 ps to 1.26 ps when cooling down the device, which corresponds to a total variation of about 170%. The latter directly transforms into a larger carrier-to-photon lifetime ratio. Assuming the photon lifetime is slowly varying compared to the upper state lifetime between 290 and 77 K [96], the carrier-to-photon lifetime is thus expected to increase from 0.13 to 0.26 on average. This partly explains the increased sensitivity to external optical feedback of the QCL, as already underlined in the Lang and Kobayashi model for semiconductor lasers under EOF in which the carrier-to-photon lifetime ratio plays a predominant role. However, in comparison with diode lasers in which the threshold leading to instabilities usually decreases at higher temperature due to a reduced output power and a larger linewidth enhancement factor [97], this result shows that the QCL becomes more sensitive to optical feedback at low temperature, despite a larger output power. As a consequence of that, other parameters such as the damping rate which was not studied in that experiment, may also vary with temperature and further influence the QCL sensitivity to optical feedback.

3.3.3 Continuous Wave

For the experiments with a pure continuous bias, a QCL from the same batch as the one used for the quasi-continuous wave experiments was utilized. This means that the two lasers are made with the same epitaxial stacks, but tiny defects in the growing process make them slightly different: the previous one can only be pumped with a 60% duty cycle at room temperature whereas the laser we are going to focus on now can be pumped with a continuous bias. Figures 3.25a and 3.26a show the evolution of the time traces when increasing the feedback strength at two different Peltier-cooled temperatures, namely 249 and 263 K, respectively, where several non-linear patterns can be experimentally observed. The LFF pattern is better visualized in Fig. 3.25a while Fig. 3.26a displays a restabilization. The changes in the retrieved waveforms might be due to the intrinsic definition of chaotic behavior for which a tiny change in the initial conditions can lead to very different results. For instance, the transition between oscillations and the LFF pattern can be clearly seen in Fig. 3.25a: some dropouts start appearing for a feedback ratio of 6.27% and then become the preponderant pattern above 6.50% (compare Fig. 3.25a plot b with plot c). In the case

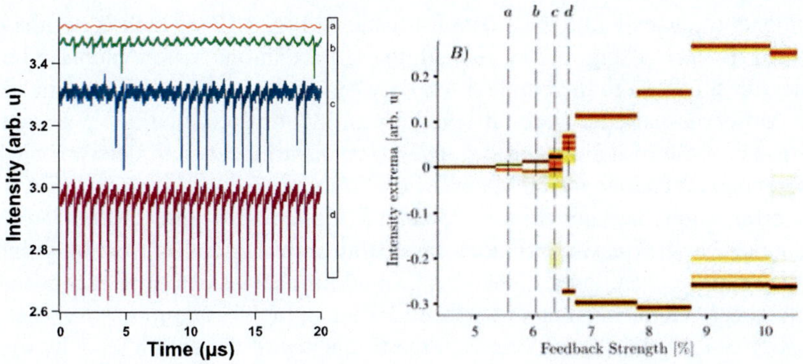

(a) (b)

Fig. 3.25 Experimental time traces in continuous mode **a** for several values of the feedback strength in % and related bifurcation diagram **b** when the temperature is 249 K. (**a**) 5.81%, (**b**) 6.27%, (**c**) 6.50%, (**d**) 6.72%; The signal is primarily steady and then evolves to non-linear patterns such as oscillations and LFFs. Traces **b** and **c** show the transition between an oscillatory state and the LFF regime, when the periodic pattern starts displaying dropouts with high amplitude. For feedback strength below 5%, the QCL's signal is stable

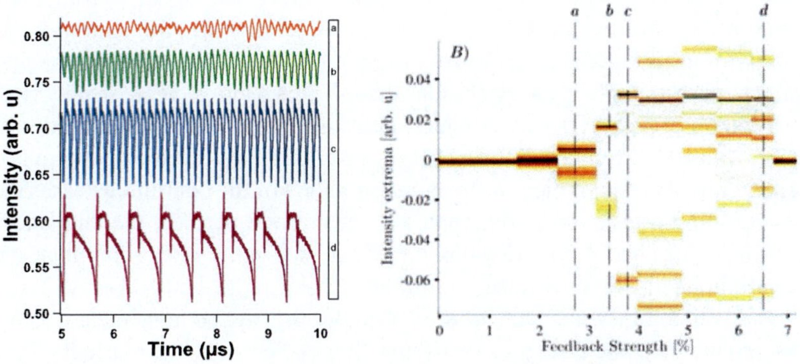

(a) (b)

Fig. 3.26 Experimental time traces for continuous operation **a** for several values of the feedback strength in% and related bifurcation diagram **b** when the temperature is 263 K. **a** 3.14%, **b** 3.56%, **c** 3.99%, **d** 6.72%; values above 7.16% lead to a stable signal and this corresponds to a restabilization, which is never seen in the case of a quasi-continuous bias. Traces (**a**) and (**b**) exhibit several extrema and could be of prime interest in the case of secure communications through chaos synchronization

of Fig. 3.26a, the transition does not look the same. The amplitude of the oscillations increases to give a chaotic pattern with several maxima and minima (Fig. 3.26a, plots c) and d)) before an homogenization of the extrema and the advent of a pattern similar to the LFF dynamics already studied. It is also worth noting that the LFF pattern in the case of Fig. 3.26a has the same curvature than the LFF pattern studied in the early experiments of chaos with QCLs (see for instance Fig. 3.17), which is not the case of the LFF pattern of Fig. 3.25a. Indeed, the latter exhibits a slow increase before a sudden drop related to the competition between modes. Albeit different in shape, the two observed patterns were retrieved for an identical configuration, except the temperature of the heat sink. Further investigations are needed to determine which parameters could change the curvature of the LFF.

The bifurcation diagrams (Figs. 3.25b and 3.26b) are also quite different from the ones obtained in the quasi-continuous case. Both show a Hopf bifurcation point but then, the route to chaos only displays a limited number of extrema, except for the transition between the oscillatory and the LFF state. In the continuous bias case, we have never observed fast oscillations at the frequency of the external cavity even if the injection current is highly above threshold ($J = 1.7 \times J_{th}$ for the continuous bias experiments). For one of the two configurations, we were able to see a restabilization for feedback ratios larger than 7.2%, until the maximum achievable feedback ratio of 10.5%. Depicted time traces in Fig. 3.26a also display, in the middle of each LFF pattern, a singular dip for the highest values of feedback ratio (see Fig. 3.26a, plot d)). This leads to another splitting in the bifurcation diagram because in that case, not only one minimum and one maximum exist. For feedback ratios between 3.99 and 6.72%, the global pattern is very similar and the repetition frequency of this pattern decreases from 4 MHz to 1.5 MHz, before the signal becomes stable again with the restabilization shown in Fig. 3.26b. In addition, the pattern with the lowest repetition frequency tends to become periodic and might not be considered as a LFF pattern though it has a similar shape. This kind of regular patterns has already been observed in QD-lasers under optical injection [98], with a similar long-timescale. Another difference between the quasi-continuous and the continuous case is the characteristic time of the non-linear dynamics. For the continuous case (which is equivalent to a quasi-continuous bias with an infinite period), the characteristic time is around 1 μs and it goes down below 0.1 μs when the laser is pumped with a quasi-continuous bias with a pulse width of 2 μs. For pulse widths below 1 μs, the QCL remains in a transient state and it is difficult to study the non-linear dynamics. This discrepancy may be due to a sum of non-linear, electrical and thermal effects in the quasi-continuous case. We will elaborate this hypothesis in section 5.2.7 dealing with extreme events in QCLs

3.4 The Entrainment Phenomenon

3.4.1 *Pioneers*

Because two different behaviors were exhibited when studying QCLs with continuous bias and quasi-continuous bias, it was relevant to further tune the bias current of the QCL to see how the non-linear dynamics evolve. We decided to focus on a sine-wave modulation and a square-wave modulation because these schemes have already been extensively studied in the case of laser diodes and provide a framework in order to draw comparisons. However, fancier modulation schemes have been tested with other semiconductor lasers [99, 100] but will not be described in this dissertation. As the LFF dynamics is highly excitable, prior works have reported on the impact of current modulation and periodic forcing on the chaotic patterns, as shown in Fig. 3.27. That can be meaningful for manipulating the phase-space dynamics with a view toward an increased control of the spiking dynamics related to LFFs, as discussed in the next section. For instance, several numerical [101–103] and experimental [104, 105] efforts have widely described the entrainment phenomenon when periodically modulating the laser current and with various schemes of optical feedback, such as strong reinjection [106] or phase-conjugate feedback [107]. However, most of them focus on modulation frequencies of several dozens or hundreds of MHz in order to have a forcing frequency close to the external-cavity frequency. Consequently, a limited number of papers deals with low-modulation frequencies below 5 MHz [108–110]. Entrainment phenomenon corresponds to a shift of the frequency of an oscillator in order to synchronize to that of a periodic forcing. Such work has been widely described in electrical, mechanical and biological systems [111]. Indeed, in the latter, triggered spikes are the main means of communication between firing neu-

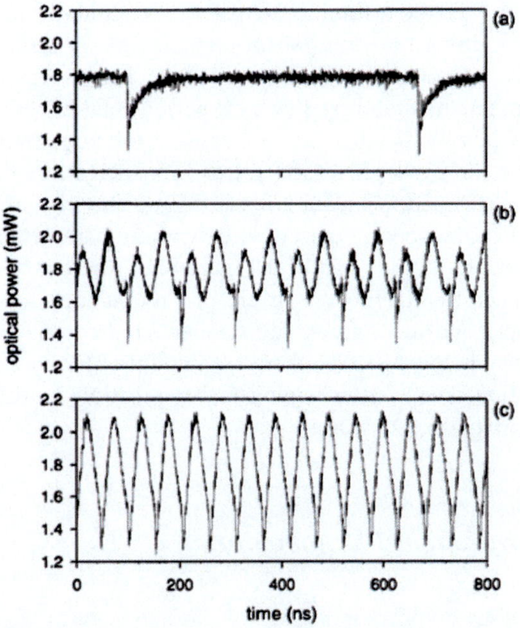

Fig. 3.27 One of the first experimental proofs of the entrainment phenomenon in a laser diode with EOF and bias current modulation. **a** LFFs in the already described case where no modulation is applied, **b** and **c** synchronization of the LFF dropouts with a 19 MHz current modulation. The time-interval between the dropouts is mostly constant, but depends on the amplitude of the sine forcing, which is 8% of the continuous bias current in (**b**) and 13.8% in (**c**). Courtesy of [116]

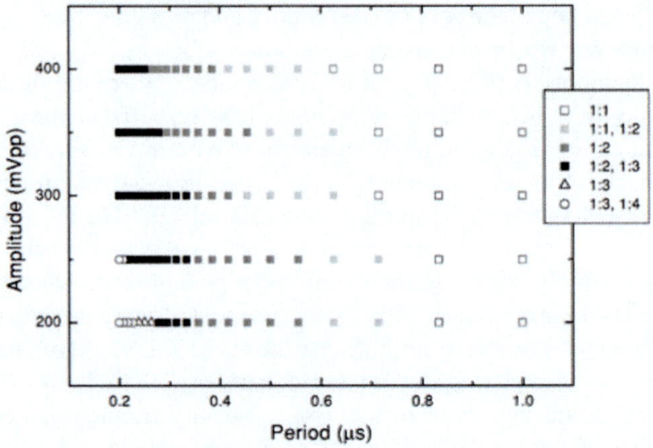

Fig. 3.28 Experimental map of the evolution of the $q : p$ criterion for a low-frequency modulation of the pump current in the case of a laser diode. The modulation is characterized by its voltage amplitude and its period. Courtesy of [114]

rons in the neuronal network [112, 113], and semiconductor lasers operating under the LFF regime and a periodic forcing are relevant candidates to mimic neuronal activities. When describing the entrainment phenomenon in semiconductor lasers, one of the key parameters is the $q : p$ criterion [114] where q represents the number of dropouts occurring in the time-series every p periods of the forcing. This criterion profoundly varies when modifying the frequency of the periodic forcing, whereas the amplitude of the forcing seems to have less influence on this criterion in laser diodes [115], unless it is widely increased [116]. Figures 3.28 and 3.29 give details about the evolution of the spiking period when varying parameters of interest in a semiconductor laser. On the one hand for Figure 3.28, there is at least one period of modulation between two LFF bursts, even for a modulation frequency as low as 1 MHz. On the other hand for Figure 3.29, the behavior is a bit different and for a modulation frequency below 10 MHz, there are multiple LFF bursts for each period of the sine modulation. These latter findings are compatible with our experimental results in QCLs, as explained hereafter.

3.4.2 The Entrainment Phenomenon in QCLs

We showed that QCLs could output LFFs when subjected to EOF, either with a continuous or a quasi-continuous bias. If the laser is pumped with a continuous bias, the time-interval between the LFF spikes is mostly unpredictable, even though recent efforts have used a symbolic method of time-series analysis in order to better predict when LFF dropouts occur [117, 118]. When, in a second step, periodic forcing is added, the time-interval between LFF spikes seems to remain constant, as can be

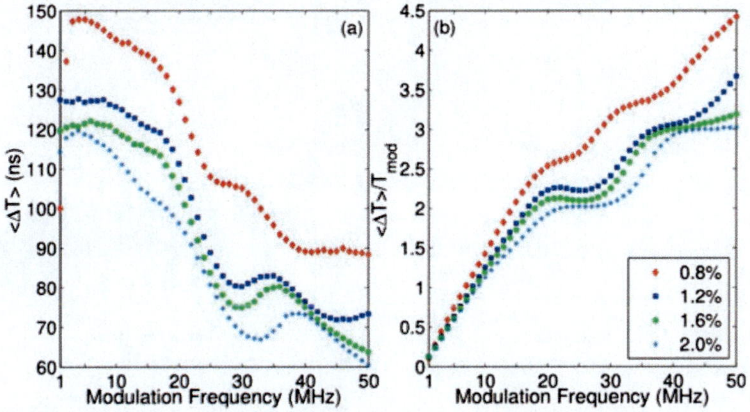

Fig. 3.29 Evolution of the mean time-interval (ΔT) between LFF spikes (left) and evolution of the ratio between the mean time-interval and the modulation frequency ($\frac{\Delta T}{T_{mod}}$) for several forcing frequencies and amplitudes of modulation (right). The latter are given in percentage and are related to the bias current of the laser diode, set at 29 mA. Courtesy of [115]

seen in Fig. 3.30 when a continuous bias of 430 mA and a peak-to-peak sine forcing of 120 mA at 2 MHz are applied. To confirm the chaotic behavior of the waveforms with EOF and external periodic forcing, we carried out a LEs analysis from the experimental time traces. Figure 3.32a, b show the phase diagram corresponding to Fig. 3.31a, b, respectively, and reveal an attractor behavior in the case where both EOF and external periodic forcing are applied. Furthermore, Fig. 3.32c emphasizes that the largest LE converges towards a strictly positive value when the EOF is applied, contrary to what happens without feedback.

In the case of the entrainment phenomenon, applying a periodic electrical forcing favors specific frequencies. As can be seen in Fig. 3.33 from the RF spectrum analysis for various continuous biases, these frequencies are mostly integral multiples of the modulation frequency, even if this relationship seems to slowly vanish when increasing the modulation frequency above 3 MHz. This may be due to the cut-off frequency of the current source which is optimized for a sine modulation up to 3 MHz. Figure 3.33a gathers all the frequencies which can be found in the RF spectrum analysis of the QCL under EOF and periodic forcing when the frequency of this periodic forcing takes values between 1.6 and 3.4 MHz in steps of 0.2 MHz. Each marker represents a frequency with a retrieved intensity above -70 dB in the case where the QCL is pumped with a continuous bias of 430 mA and a modulation amplitude of 120 mA. Most of the markers appear on one of the solid lines representing integer multiples of the modulation frequency, which means that the time-interval between successive LFF spikes strongly depends on the forcing frequency. The same conclusion can be drawn for Fig. 3.33b, c. This relationship between the modulation frequency and the shape of the RF spectrum had already been revealed in the case of laser diodes, for higher frequency modulations [105]. Figure 3.34 shows the time series, phase diagrams and electric spectra in the case of a modulation frequency of

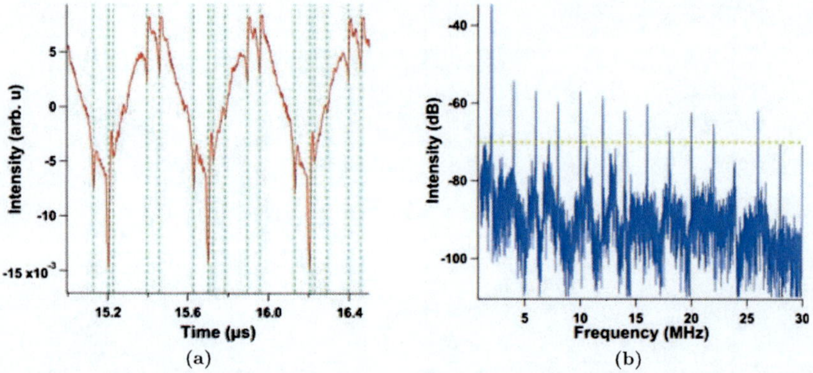

Fig. 3.30 Experimental time trace close-up (**a**) and RF spectrum (**b**) of the QCL's emitted wave when a periodic forcing at 2 MHz is applied with external optical feedback; vertical dashed lines represent the dropout occurrences in the time trace and same figure without these lines can be found in Fig. 3.31c; the interval between two spikes in the RF spectrum is exactly the value of modulation frequency $f_m = 2\,\text{MHz}$ and the horizontal dashed line represents the intensity threshold taken into account for Fig. 3.33

500 MHz for two levels of modulation strength. The aforementioned phenomenon is more significant in the case of a high-amplitude sine modulation and Fig. 3.34f can be accordingly compared with Figure 3.30b. Even if the RF spectrum seems to globally show the same behavior whatever the modulation frequency, the time traces are quite different in terms of the distribution of the spikes, as shown in Fig. 3.31. This behavior can be studied using the aforementioned $q : p$ criterion. In our case, the repetition frequency of the spikes is always higher than or equal to the frequency of the forcing, so we set p to 1. Tables 3.2 and 3.3 gather the q values for several continuous biases and modulation amplitudes when the frequency of the forcing is 1.5 MHz and 2 MHz. These tables show that, on the one hand, q slowly varies when increasing the value of the continuous bias. On the other hand, q depends on the variations of the amplitude. To summarize: not only the frequency of the periodic forcing, but also the amplitude of the periodic forcing has a preponderant influence on the time-interval between spikes. Figure 3.31 shows that the q parameter deeply varies for the range of frequencies under study. For instance, q equals 1 for a modulation frequency of 3.4 MHz, and 6 for a modulation frequency of 2 MHz when the QCL is pumped with a 430 mA bias and when the amplitude of the modulation is 120 mA. From 1.5 to 2.7 MHz, each period of the sine wave gathers multiple LFF spikes and it is difficult to forecast a specific organization of the LFF dropoffs. When the modulation frequency is above 2.7 MHz, only two LFF patterns occur per wave period and the dropoffs always occur for a fixed phase of the periodic forcing, which is when the extrema of the sine wave are reached. Eventually, for frequencies above 3.4 MHz, the repetition frequency of the spikes is exactly the same as the forcing frequency (Fig. 3.31f) or, in other words, q equals 1. However, as mentioned in the section describing the experimental setup, the low-noise current source has a cut-off

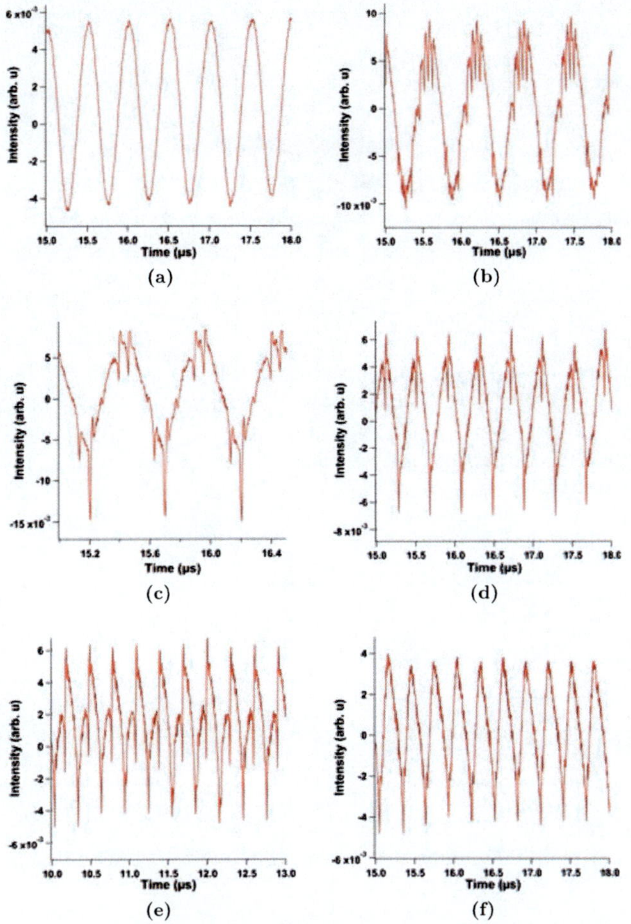

Fig. 3.31 Experimental time traces when EOF is applied to the QCL and with a sine forcing of the continuous wave of: 1.6 MHz (**b**), 2 MHz (**c**), 2.7 MHz (**d**), 3.3 MHz (**e**), 3.4 MHz (**f**); trace **a** corresponds to the case when no EOF is applied and the modulation frequency is 2 MHz. All traces were retrieved for a continuous bias of 430 mA and a modulation amplitude of 120 mA

frequency around 3 MHz and even if the sine modulation is still visible in the time traces with no EOF, the amplitude of the forcing is less important compared to the modulations with frequency below 3 MHz. It is worth noting that in the case where the period of the dropoffs and the period of the sine forcing are the same, or q equals 1, the spikes appear for a given phase, as previously described. Nevertheless, the value of this phase can vary for two different configurations. For instance, Fig. 3.35 shows that the phase-shift is π between the two configurations, even though the value of q remains 1. Further investigation is needed to determine what can make spikes occur simultaneously with maxima or minima of the sine forcing.

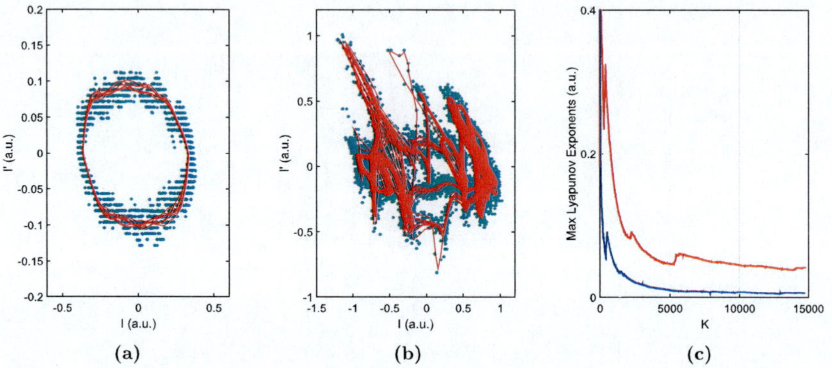

Fig. 3.32 Dynamics analysis of the experimental waveforms. Phase portrait corresponding **a** to Figs. 3.31a, b to 3.31b. Blue dots are the measured raw data and the solid red curves represent the noise-reduced orbital trajectories. **c** gives the calculated largest LE related to Fig. 3.31a (blue curve) and to Fig. 3.31b (red curve)

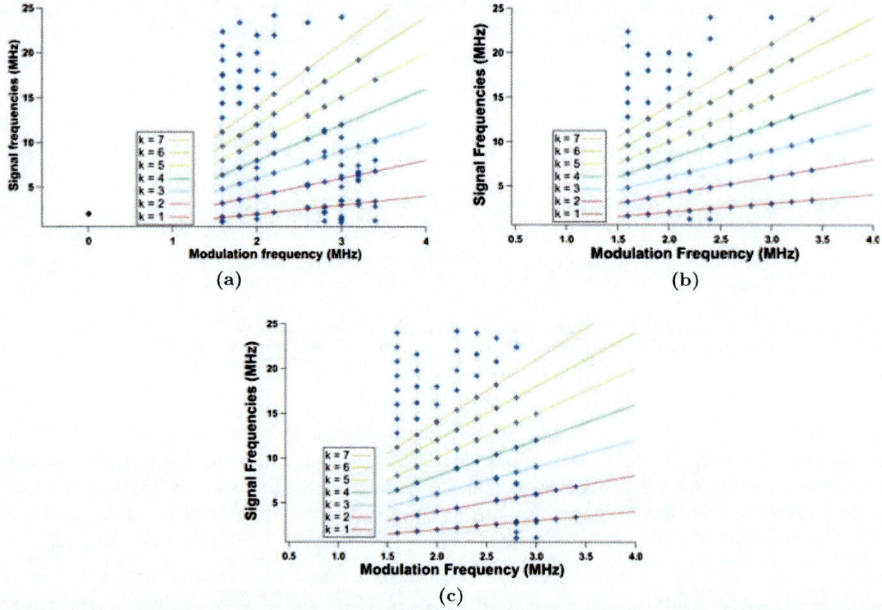

Fig. 3.33 **a** RF spectrum of the chaotic signal when applying a 120 mA peak-to-peak sine periodic forcing to the QCL under a continuous bias of 430 mA; blue markers correspond to the RF spectrum frequencies above −70 dB, when EOF is applied for modulation frequencies shown on the x axis while the black marker represents the RF spectrum of the signal when a 2 MHz modulation is applied without EOF; solid lines are for visual guidance for the reader and represent integral multiples of the modulation frequency: $f_s = k \times f_m$ with k displayed on the diagram. Figure **b** and **c** correspond to the same configuration but for a continuous bias of 350 and 530 mA, respectively

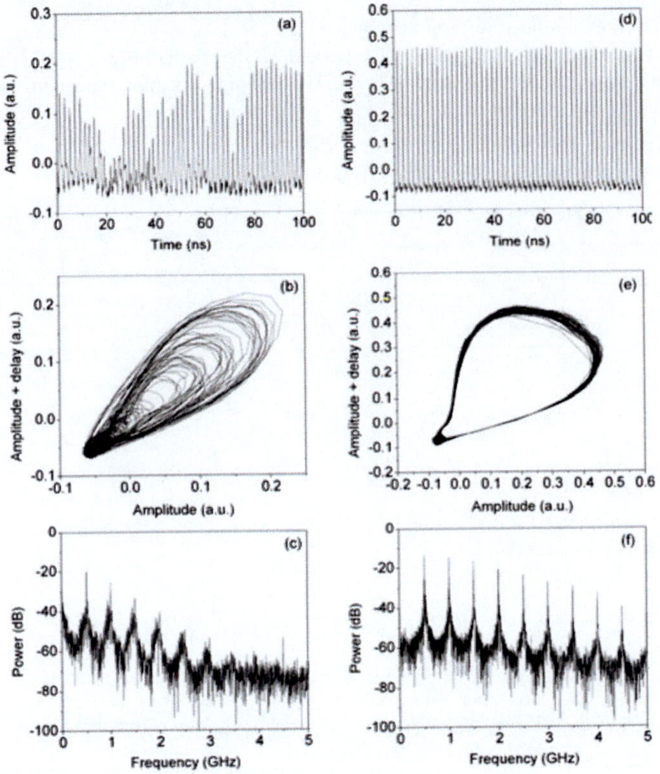

Fig. 3.34 Experimental results in the case of a laser diode subject to the entrainment phenomenon at high frequencies. **a** Experimental output power time-series, **b** phase portrait and **c** frequency spectrum for a modulation amplitude of 0.4 V. **d**, **e** and **f** same plots but for a modulation amplitude of 1.25 V. Courtesy of [105]

Table 3.2 q value with related uncertainties when the frequency of the sine forcing is 1.5 MHz. For high q values, the significance is due to the irregular spiking pattern that resembles LFF phenomenon

Modulation amplitude\ Continuous bias	350 mA	430 mA	530 mA
40 mA	3	4	4
80 mA	7 ± 1	8 ± 1	8 ± 1
120 mA	10 ± 2	10 ± 1	12 ± 1
160 mA	14 ± 2	14 ± 2	15 ± 1

Most of the studies focus on a sine-wave modulation that is applied to the laser, the latter being moreover pumped with a continuous current above threshold. Here, we also study the case of a square-wave modulation. A square-wave modulation is not identical to a quasi-continuous wave in the sense that a continuous bias is

Table 3.3 q value with related uncertainties when the frequency of the sine forcing is 2 MHz; the value with the dagger corresponds to the time trace and RF spectrum of Fig. 3.30 and the value with the asterisk corresponds to the time trace of Fig. 3.35a. For high q values, the significance is due to the irregular spiking pattern that resembles LFF phenomenon

Modulation amplitude\continuous bias	350 mA	430 mA	530 mA
40 mA	1*	2	2
80 mA	3	5	5 ± 1
120 mA	6 ± 1	$6 \pm 1^{\dagger}$	8 ± 1
160 mA	8 ± 1	9 ± 1	9 ± 1

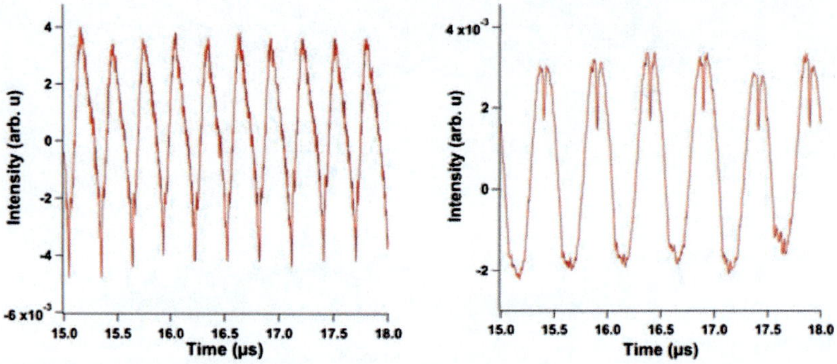

Fig. 3.35 Experimental time traces for a continuous bias of 350 mA and a modulation amplitude of 40 mA (**a**) and for a continuous bias of 430 mA and a modulation amplitude of 120 mA (**b**); both experimental waveforms illustrate the condition q equals 1

applied prior to the addition of the square-wave modulation. A quasi-continuous bias starts well below threshold and goes up to the required value of bias current. Most of the time, the starting point is at a few dozens of mA because when remaining too close to 0 mA, the voltage is quite unstable and this can also lead to a voltage slope that is beyond the specifications of the current source, especially in a cryogenic environment where LIV characteristics are highly sensitive to temperature [119]. If the periodic forcing is not a sine wave but a square wave, like shown in Fig. 3.36a, the conclusions remain similar for the RF spectrum analysis. Figure 3.36b shows that the frequency interval is 0.5 MHz when the bias forcing is a square wave with a frequency of 0.5 MHz and a duty cycle of 90%. The time trace also displays a synchronization of the LFF dropouts as seen in Fig. 3.36a. Dropouts always appear for a given phase-shift of the square signal, especially along the slopes of the square. But contrary to the case with a sine-wave modulation, the time-interval between the dropouts is not constant anymore. Dropouts are mainly gathered at the upper edge of the square and are quite rare in the plateau region corresponding to the constant value of the upper part of the square. It is worth noting that despite the similarities of

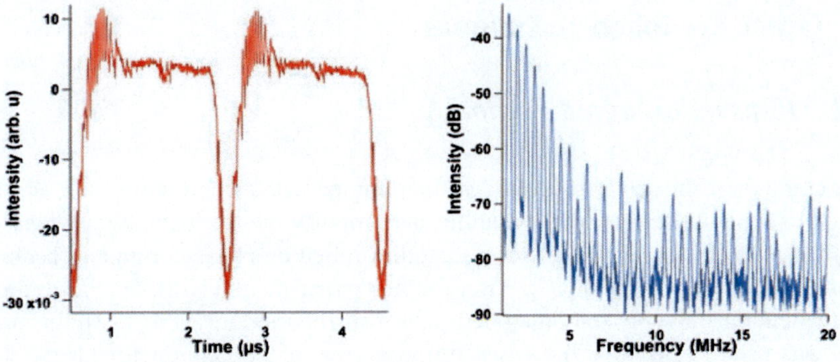

(a) (b)

Fig. 3.36 Experimental time trace (**a**) and RF spectrum (**b**) when EOF and a square-wave modulation at 0.5 MHz with an amplitude of 28% are applied

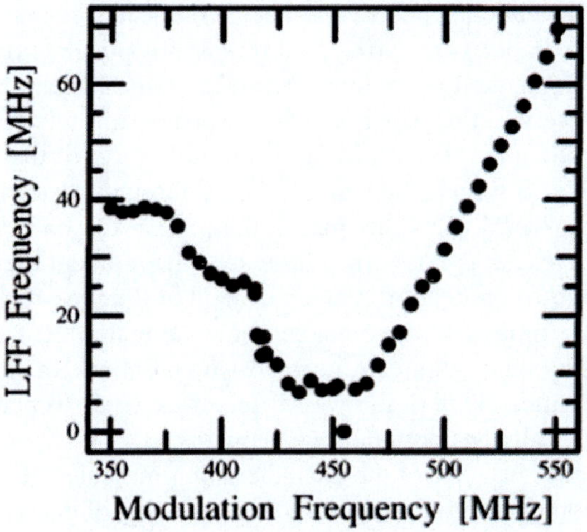

Fig. 3.37 Average repetition frequency of LFFs for high modulation frequencies of the pump current in the case of a laser diode. For a forcing frequency matching that of the external cavity (455 MHz), the LFF phenomenon is totally suppressed. Courtesy of [120]

the RF spectra in both modulation cases, the dropouts synchronize with the forcing in several ways. This observation could lead to a very precise triggering of such dropouts and can be highly relevant for secure communications.

Further investigation will determine the influence of the shape of the modulation (square signal, asymmetric triangular pulses or sine forcing) as well as the evolution of the $q : p$ ratio when increasing the modulation frequency. The external-cavity frequency is actually set to be of a few hundreds of MHz. Experimental studies with laser diodes which achieved to match the frequency of the forcing with the frequency of the external cavity showed the absence of LFF dropouts for a narrow range of modulations [120], as exposed in Fig. 3.37. It is also relevant to note that the studied range of modulation frequencies did not allow us exhibiting ghost frequencies [121] in the RF spectrum analysis. Indeed, all the retrieved frequencies, which include the frequencies of the LFF spikes, were at least equal to that of the forcing.

3.5 Other Feedback Techniques

3.5.1 Phase-Conjugate Feedback

In the case where the semiconductor laser is pumped with a continuous bias, another method to expand the observed dynamics is to modify the feedback light when it is travelling in the external cavity. Several studies relied on phase-conjugate feedback instead of conventional feedback. After the first experimental work on a laser diode with feedback from a phase-conjugate mirror was reported in 1985 [122], this technique was used to improve the emission properties of semiconductor lasers, with applications similar to those described in a further chapter, in the case of optical injection. Phase- and mode-locking [123], frequency stabilization [124], spectral linewidth [125] and intensity noise improvements [126] were demonstrated.

On the one hand, the main advantage of phase-conjugate feedback compared to conventional feedback is that the reinjected beam is self-correcting, self-aligning and aberrations are corrected through the optical components [127]. On the other hand, this method is far more complex to implement, as detailed below. Self-correcting describes the principle that any phase-shift and spreading that the laser beam could undergo in its path from the emitting facet of the semiconductor laser to the mirror due to propagation in the air or through distorting media is compensated in the backward trip. This means that, when fed back into the laser cavity, the photons have exactly the same phase as at the moment they were emitted out of it. Actually, phase conjugation is an equivalent of the time-reversal phenomenon [128], which is not limited to the optoelectronics domain [129, 130]. Once the beam has reached the phase-conjugate mirror, it travels back in time until the instant when it was emitted. This time-reversal feedback leads to peculiar non-linear dynamics which are not always found in a semiconductor lasers with conventional optical feedback [131] because of the characteristic timescales of the crystals used to achieve phase-conjugate feedback [8]. Indeed, the order of magnitude can range from microseconds (in tellurium-doped tin hypothiodiphosphate) to several seconds (in barium titanate crystals) whereas the typical carrier response in semiconductor lasers with optical feedback is of the order of one nanosecond, defined by the laser relaxation oscillation frequency. Therefore, typical effects of phase-conjugate feedback occur when the phase-conjugate mirrors respond as fast as this timescale. The principle of phase conjugation with a photorefractive crystal is based on four-wave mixing [132]. The interferences between two pump beams result in the generation of a refractive index grating within the crystal. The beam from the semiconductor laser is sent towards this grating and the resulting scattering on the grating creates a fourth beam which properties are to be counter-propagating and phase-conjugate with respect to the initial beam.

In terms of self-pulsing dynamics, the bifurcation mechanism may be different in the case of EOF compared to the case of phase-conjugate feedback. A pioneering theoretical analysis of the laser dynamics and bifurcations with phase-conjugate feedback has been performed by Krauskopf and co-workers in 1998 [133] who depicted an in-depth account of the evolution of the bifurcation diagram and the route to chaos. With EOF, the bifurcation process is a Hopf bifurcation on an external-cavity mode, which is a steady-state solution of the system, with a competition between the external-cavity modes and the laser cavity modes. The self-pulsing dynamics thus corresponds to a bridge of periodic solutions connecting two Hopf bifurcations related to the aforementioned modes [134]. However, one of the bridging modes is actually an unstable solution, or anti-mode, so this is the reason why the observed non-linear dynamics are easily destabilized to deterministic chaos when increasing the feedback strength. This makes the observation of the beat mode and the Hopf bridge quite difficult. In the phase-conjugate feedback case, the laser has only one stable steady state solution that is destabilized through a Hopf bifurcation to a stable self-pulsing solution. The frequency of this solution is close to that of the external cavity, that solution has been called external-cavity mode but by contrast to EOF, an external-cavity mode of the phase-conjugate feedback system is a rotating solution with time-periodic intensity and not a time-constant intensity. When the feedback strength increases, the laser creates new external-cavity mode solutions with the same origin for all of them i.e. a single steady state branch. This is not the case for the EOF, because a possible beating or Hopf bridge on external-cavity modes can occur. The stability of the external-cavity self-pulsing dynamics in the phase-conjugate feedback case is determined by the interplay between a saddle-node bifurcation and a torus bifurcation. Consequently, it is possible to observe a robust and fully stable self-pulsing also when increasing the feedback rate or the external-cavity length, which is not compatible with the several non-linear regimes studied in Sect. 3.2.5. Phase-conjugate feedback has also been recently investigated in laser diodes in order to trigger chaos with a bandwidth larger than that set by the relaxation frequency [135, 136].

3.5.2 *Rotated Polarization Feedback*

Semiconductor lasers usually emit linearly polarized light due to the symmetry of their gain medium and their inherent material anisotropy. This is typically the case for optical sources relying on intersubband transitions [137] because only the component of the electric field normal to the quantum wells can optically couple to the quantum oscillators, which results from quantization of the allowed displacement. The parallel components are left to interact weakly with a free two-dimensional electron gas. For a symmetric quantum well, the ratio $\overline{r}$ between the two orthogonal components is given by [138]:

$$\overline{r} = \zeta \frac{(E_{n'} - E_n)\Delta_0}{(E_n + E_g + \Delta_0)(E_{n'} + E_g)} \tag{3.31}$$

with $\zeta < 1$ a constant coefficient depending on the subband index difference parity, $(E_{n'} - E_n)$ the energy separation between bands, Δ_0 the spin-orbit splitting and E_g the fundamental energy gap of the material. Typical theoretical values are around

1% and are in good agreement with experimental studies [139]. For instance, QCLs emit a wave which is mostly TM polarized, even if recent efforts showed it is not perfectly linearly polarized, and more specifically for the DFB-QCL case [140]. Among semiconductor lasers, VCSELs stand as an exception. Because of the cylindrical symmetry of their resonator, the isotropy of their gain and the polarization independent reflectivity of their Bragg mirrors, they have a priori no preferred polarization [141], and several methods have been developed in order to switch the polarization on demand [142, 143]. Other configurations, such as pumped Er^{3+}-doped-fiber-ring laser (EDFRL) can also exhibit TM and TE polarizations and were studied as testbeds for polarization switching dynamics, both experimentally and numerically [144], as shown in Fig. 3.38. At first sight, polarization switching can be detrimental in optical communications but when thoroughly controlled, it is of prime interest in optical routing [145], clock recovery [146] and random number generation [147]. When not controlled, the switching between polarization can be triggered by noise or inherent to chaotic dynamics [148]. On the other hand, periodic switching can be achieved through optical pulse pumping [149], optical injection [150] and with rotated polarization feedback [151]. The latter configuration is of paramount importance when it comes to study the possibility of polarization switching in QCLs, because these experiments were carried out with laser diodes. Contrary to VCSELs, laser diodes exhibit single TE mode emission with high TM mode suppression when lasing as a solitary laser without EOF, and this is similar to what can be found in QCLs (with the TM polarization being greatly favored in the case of QCLs). In order to rotate the polarization of a wave before feeding it back to the laser, there are three possibilities. The first one is to insert a quarter-wave plate (QWP) within the external cavity. If the light is perfectly TM polarized (as it is almost the case in QCLs), the polariza-

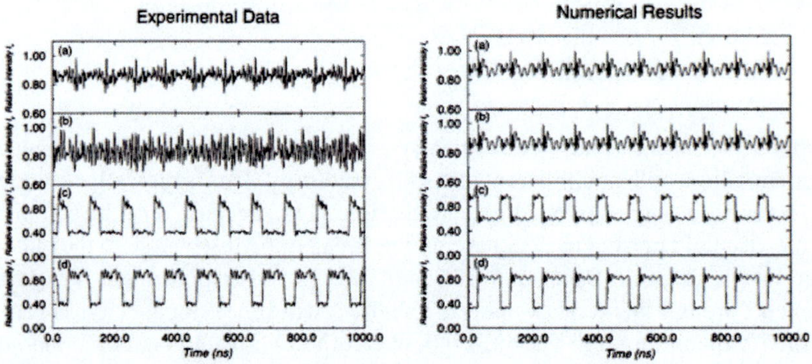

(a) (b)

Fig. 3.38 Experimentally measured polarization intensity (left) and numerical simulations (right) of **a** self-pulsing in the x-polarization direction from an EDFRL with 10% output coupling, **b** self-pulsing in the y-polarization direction. The EDFRL was pumped at 4 times threshold. **c, d** Antiphase square pulses in the x- and y-polarization directions, respectively, from an EDFRL with 3% output coupling. The EDFRL was pumped at 3.3 times the threshold current. Courtesy of [144] tion of the wave will become circular after the first travel through the QWP. Then, the wave is reflected by the mirror, so the polarization remains circular but in the opposite direction. On its way back to the laser, this circularly polarized wave travels once again through the QWP, which converts it into a TE wave. Consequently, this methods allows turning a linearly polarized wave into a linear orthogonal wave, as illustrated in Fig. 3.42. However, if the initial wave is not purely linear, the feedback wave will have a polarization more complex than the one previously described, and it is difficult to control what is the exact amount of TM/TE polarization traveling back to the laser. Two other methods have been described in order to select only one linear polarization for the reinjected wave. The first one is a loop-injection with an optical isolator, a half-wave plate and a polarizer [152]. The second one replaces the QWP by a Faraday rotator and a polarizer, so that the initial wave is first rotated by 45°, then polarized along the direction given by the polarizer before being reflected by the mirror of the external cavity and eventually, rotated again when coming back through the Faraday rotator [153]. Both of these methods are designed to avoid multiple round-trip reflections and allow only unidirectional coupling between the natural horizontal polarization mode and the unsupported orthogonal mode. Yet, the latter method is the most studied because Faraday rotators in the near-infrared are commonplace.

With polarization rotated feedback, it is possible to observe chaos and square-wave polarization switching, even in the case of laser diodes where the TM polarization is almost suppressed. This square-wave pattern appears for high feedback ratios, and the frequency of the pattern is set by the frequency of the external cavity [151], as visualized in Fig. 3.39. It was also shown that the duty cycle of the square can be other than 50%. In a scheme with two laser diodes shown in Fig. 3.40, Sukow and co-workers successfully tuned the duty cycle by varying the bias current of one of the two lasers while the other one was kept constant [154]. Even when the duty cycle is not 50%, the switching between the TE mode and the TM mode occurs, which means that as soon as the wave is suppressed in the TE mode, the TM mode

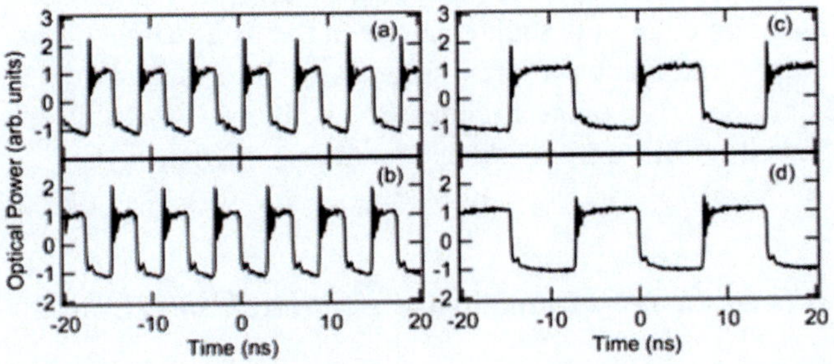

Fig. 3.39 Experimental time series of the square-wave self-modulated oscillations which were obtained in the case of a laser diode subject to cross-polarization external feedback. Top and bottom, TE and TM mode intensities, respectively. The left panel corresponds to the case where the external-cavity round-trip time is 2.60 ns, and the right panel corresponds to the case where the external-cavity round-trip time is 6.92 ns. Courtesy of [151]

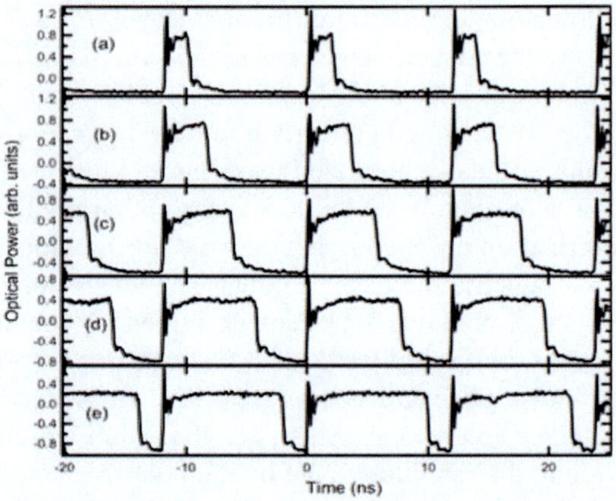

Fig. 3.40 Experimental time series of the square-wave self-modulated oscillations which were obtained in the case of a laser diode subject to cross-polarization external feedback. The duty cycle of the square pattern varies as a function of pump current. Only the TE mode of the laser diode is illustrated for clarity, and the TM mode is the symmetry of the TE mode, as shown in Fig. 3.39. Courtesy of [154]

is triggered, and vice versa. More complex dynamics can be found, depending on the feedback strength, while the switching between the TM polarization and the TE polarization remains [155]. In VCSELs, similar phenomena can be observed, which is not surprising due to the existence of both polarizations. Recent experiments and simulations showed that the square-wave switching was only possible for high cross-polarization feedback strength [156]. For values below 30%, the VCSEL emits first undamped relaxation oscillations and then chaos. For feedback strengths above 30%, the VCSEL outputs the aforementioned square pattern with a period equal to that of the cavity round-trip time. Nevertheless, an undamped oscillation is displayed on top of the square pattern with a frequency higher than the relaxation frequency of the laser under study. Further increase of the feedback strength makes these oscillations disappear while maintaining the square pattern at the external-cavity frequency and eventually, a very large feedback strength gives birth to more complex dynamics, similar to the ones observed in laser diodes [155]. Figure 3.41 details the non-linear dynamics observed for the aforementioned feedback strength.

3.5.3 Effects of Cross-Polarization Feedback in QCLs

Following these results about polarization switching in several semiconductor lasers, we experimented a similar scheme with QCLs, as illustrated in Fig. 3.43. As we did not possess a Faraday rotator nor a half-wave plate for mid-infrared wavelength, we worked with a QWP in the configuration where the feedback light is not filtered to

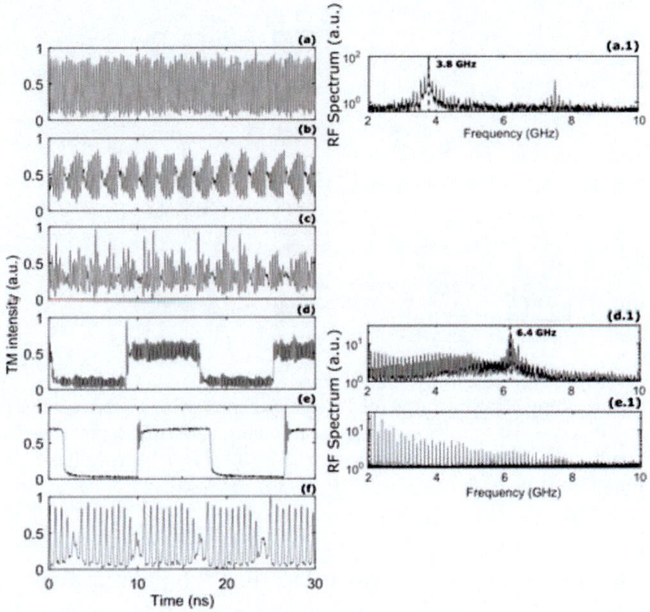

Fig. 3.41 Experimental results for a VCSEL subject to strong rotated polarization feedback. Time traces of the TM mode for **a** 20%, **b** 23%, **c** 26%, **d** 30%, **e–f** 37% feedback ratio. The feedback delay time is fixed at 8 ns. (a.1), (d.1) and (e.1) are the RF spectra corresponding to the time traces in (**a**), (**d**) and (**e**) respectively. Courtesy of [159]

Fig. 3.42 characterization of the TE light (red) and the TM light (blue) that is reinjected into the QCL after one roundtrip in the external cavity with the QWP

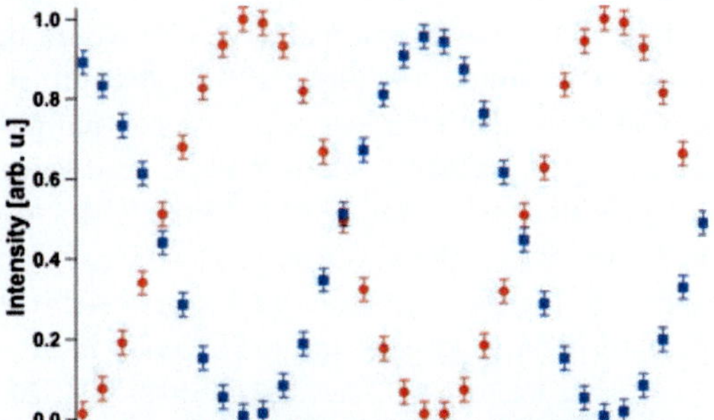

ensure a linearly polarized wave. This means that the polarization of the light coming back to the laser depends on the polarization of the emitted light in a previous outcome. We did not perform an accurate measurement of the polarization of the continuous wave generated by the QCL under study, but a recent paper using a QCL from the same manufacturer as ours, showed that there is not only a contribution of the TM mode, especially for the DFB case [140]. An analysis of the normalized Stokes parameters [157] showed that parameter S_3 is non-zero with a mean value

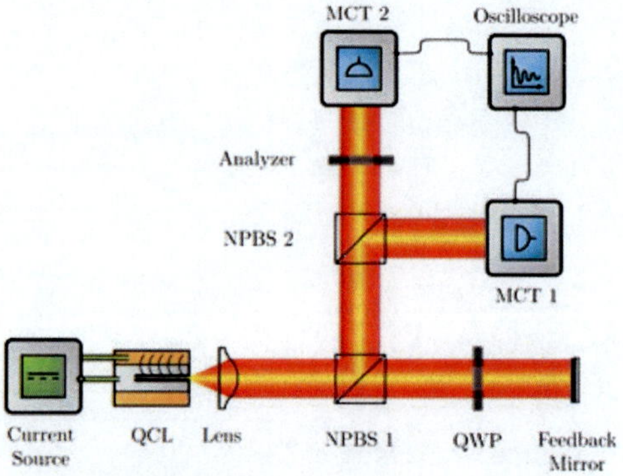

Fig. 3.43 Setup for the rotated polarization feedback experiment with a quarter-wave plate (QWP); MCT1 acts as a reference signal for phase-shift measurements, and MCT2, preceded by an analyzer polarizer, detect either the TE or the TM radiation. NPBS: non-polarizing beam-splitter, MCT: Mercury-Cadmium-Telluride

of 0.17, as visualized in Fig. 3.44. This means that the polarization is not purely linear because in this case $S_2 = S_3 = 0$. The existence of a circular polarization in the output beam can be attributed to some introduced strain [158], in the vicinity of the laser facet, by technology and processing, as mounting and soldering.

Another drawback of our system with the QWP was that we owned no variable attenuator working in the mid-infrared. The attenuator we usually consider for the EOF experiments is a polarizer, but this is not suitable for our cross-polarization scheme in a straight external cavity. It would have however been possible to use a polarizer to adjust the TE feedback strength if we had got a second QWP and if we had implemented a loop-cavity resembling that of Ref. [152]. Consequently, we have a high feedback strength with a fixed value of nearly 35% and we observe the wave emitted by the QCL, both in the TM and the TE mode. Results are shown in Fig. 3.45. When the QWP is set to 45°, we observe a square pattern for the TM mode and for the TE mode, with a phase-shift between the two, which means that when the TM mode is on, the TE mode is off, and reciprocally (Fig. 3.45a). This is very similar to what was reported for VCSELs and laser diodes. However in our case, the period of the square pattern is not related to that of the external cavity, which is surprising because this feature was usually observed whatever the type of semiconductor laser under study. In our case, the external-cavity round-trip time is 2 ns whereas the period of the square pattern is roughly 400 ns. Because we are not in the configuration with the Faraday rotator, multiple beam reflections can occur and this would lead to non-linear dynamics with periods longer than that of a single round-trip time. Yet, this possibility is very unlikely in our case because this would imply that what we observe is a pattern linked to roughly 200 round-trips within the external cavity. Furthermore, it is difficult to predict the frequency of the square pattern when changing the length of the external cavity. Changing the angle of the QWP can modify the ratio between the TE mode and the TM mode while maintaining

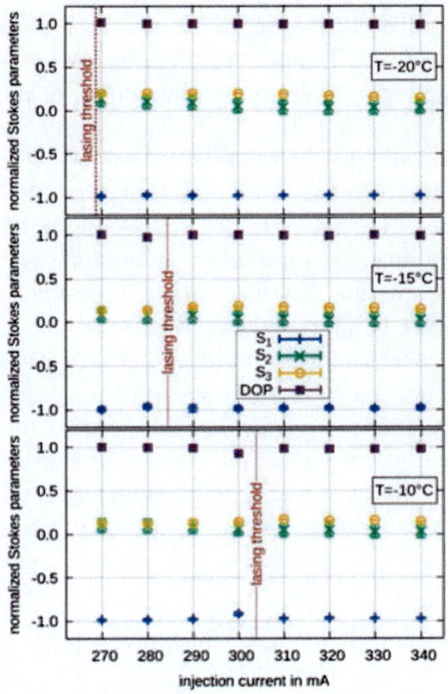

Fig. 3.44 Normalized Stokes parameters (S_1, S_2, S_3) and degree of polarization (DOP) of a DFB-QCL as a function of the injection current for three temperatures. The red dashed line indicates the corresponding threshold current. Courtesy of [140]

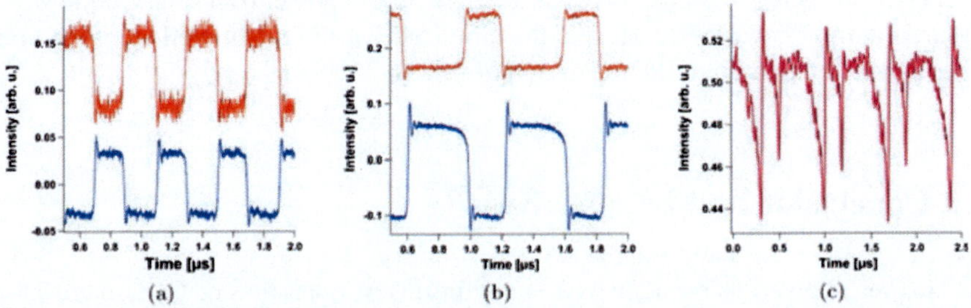

Fig. 3.45 **a** Experimental time traces of the wave emitted by the QCL under study when cross-polarization feedback is applied and the quarter-wave plate is set to 45 °. The blue waveform corresponds to the TM wave, the red waveform corresponds to the TE wave and the duty cycle in both cases is close to 50%. **b** Same configuration but for a quarter-wave plate set to 35 °. In this case, the TM wave has a duty cycle of 65% while the TE wave has a duty cycle of 35%. For these two charts, the amplitude of the TE mode is magnified to be comparable with that of the TM mode. **c** Response of the detector when $\theta = 10\,°$, showing both modulated modes with similar amplitude

the square pattern. When the QWP is set to 35°, the square pattern is split between the TM mode for 65% and the TE mode for 35%, the two modes remaining phase-shifted as shown in Fig. 3.45b. Even though the length of the external cavity was not changed, the period of the pattern is different. It is relevant to note that in the two aforementioned cases (angle of 45° and 35° for the QWP), no oscillations were spotted on top of the square pattern, as reported in some recent experiments with VCSELs [159]. Nonetheless, even in the VCSEL configuration, these oscillations, which are faster than the relaxation frequency, are not always present and strongly depend on the feedback strength. The lack of such oscillations in our experiment with QCLs can be explained either by the non-existence of relaxation frequency in QCLs or by the limited bandwidth of the MCT detector we used (estimated 3-dB bandwidth of 850 MHz per manufacturer's specifications) or by the fixed value of the feedback strength. In order to measure the ratio between the modulated TE mode and the modulated TM mode, we tilt θ, the angle of the analyzer polarizer, so that both signals are displayed in the same time trace and with the same amplitude, as illustrated in Fig. 3.45c. By comparing their amplitude, it is possible to derive the following relationship:

$$\frac{A_E}{A_M} = \frac{cos^2(90-\theta)}{cos^2(\theta)} = 0.03 \pm 0.01 \tag{3.32}$$

where $\frac{A_E}{A_M}$ is the ratio between the modulated TE mode and the modulated TM mode. In the case where $\theta = 10°$, both squares have nearly the same amplitude. Consequently, $\frac{A_E}{A_M} = 0.03 \pm 0.01$. It is worth noting that the average output power of the TE mode is less than 0.5 % of the average output power of the TM mode when measured with a powermeter. Hence, the amplitude of the modulated signal in both modes cannot be related to the mean output power.

3.6 Conclusion and Perspectives

This chapter allowed us broadening the conditions of operation of QCLs subject to EOF. A numerical analysis of the effect of temperature on the QCL's carrier lifetime gave indications about the increased sensitivity of QCLs at low temperature. This sensitivity to perturbations was also exhibited in the case of a QCW instead of a CW. However, the purpose of all these studies remains to achieve a secure transmission with chaotic QCLs and in this context, the QCL will have to be biased in CW mode in order to enable the message implementation. Moreover, a Peltier-cooled temperature will also be required to avoid the bulkiness of a vacuum cryostat even if chaos is more developed at cryogenic temperature. As QCW and CW operation are not the only options when it comes to bias a laser diode, we decided to apply a periodic modulation of the injection current.

We experimentally reported on the observation of chaos synchronization-induced optical bursts in a mid-infrared QCL. Using external optical feedback and periodic sine forcing of several MHz, the QCL operating in the LFF regime displays frequencies which are integer multiples of the forcing frequency. This differs from what was observed with only external optical feedback. This means that the distribution of the dropouts is significantly modified by the modulation of the current bias. We furthermore found that the spikes observed in the waveforms related to the LFF phenomenon occur for a given phase-shift. This phase-shift depends on the amplitude of the periodic forcing, the value of the continuous bias applied to the QCL and the frequency of the sine forcing. It also appears that the dynamics of the system are more affected by variations in the modulation amplitude than by changes in the continuous pumping bias. We thus derived that the two parameters of prime importance in order to control the number of spikes per period are the frequency and the amplitude of the modulation while the value of the continuous bias is kept constant. One particular source of applications of such experimental results concerns the field of neuronal activities and the communication between neurons due to sudden bursts, as well as other physiological processes relying on non-linear phenomena [160, 161].

We also gave, in this chapter, the first proof of a non-linear pattern in the TE mode of a QCL with rotated polarization feedback. Indeed, we previously carefully checked that absolutely no TE dynamics can be observed in a QCL with EOF. However, we need further studies to fully understand the relationship between the period of the square pattern and the polarization of the feedback light. This conclusion also concerns the duty cycle of the square pattern. With this knowledge, one can consider using QCLs in all-optical square modulation schemes, aiming at reproducing neuronal clusters. To realize such complex systems, essential building blocks, such as logic elements and modulators [162], are required for information processing. The advantage of photonic memories is to be both high-speed and energy-efficient [163]. Such experiments have already been carried out with visible and near-infrared light but the data transmission can still be increased by wavelength multiplexing, thus pushing the all-optical modulators efforts towards higher wavelengths. Concerning the low square-pattern frequency value we encountered, one indication could be found in another kind of highly-damped semiconductor lasers. Indeed, studies about optical injection in quantum dot lasers proved it was possible to trigger a square-wave pattern with a period in the order of 1 μs [164]. Both the experiments and the numerical analysis showed that only a precise wavelength detuning between the drive laser and the response laser could create a regular periodic pattern, as illustrated in Fig. 3.46. Otherwise, it is possible to trigger a square pattern but it randomly occurs and then, it is not possible to take advantage of it for applications.

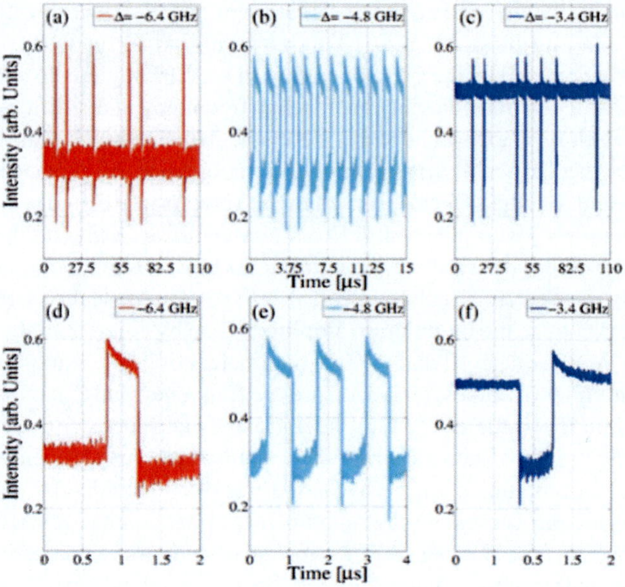

Fig. 3.46 Influence of the wavelength detuning in the case of an injected quantum dot laser. **a** and **d** randomly separated pulses at −6.4 GHz detuning. **b** and **e** periodic squares at −4.8 GHz detuning and **c** and **f** square dropouts appearing randomly at −3.4 GHz detuning. An appropriate detuning is thus able to generate a square pattern with a period close to 1 μs. Courtesy of [164]

References

1. Maiman TH, Hoskins R, d'Haenens I, Asawa CK, Evtuhov V (1961) Stimulated optical emission in fluorescent solids. II. Spectroscopy and stimulated emission in ruby. Phys Rev 123(4):1151
2. Kimura T, Otsuka K (1970) Response of a CW Nd^{3+}: YAG laser to sinusoidal cavity perturbations. IEEE J Quantum Electron 6(12):764–769
3. Maiman T (1960) Stimulated optical radiation in ruby. Nature 187:493–494
4. Mooradian A (1985) Laser linewidth. Phys Today (USA) 38(5)
5. Lorenz EN (1969) The predictability of a flow which possesses many scales of motion. Tellus 21(3):289–307
6. Strogatz SH (2018) Nonlinear dynamics and chaos: with applications to physics, biology, chemistry, and engineering. CRC Press
7. Haken H (1975) Analogy between higher instabilities in fluids and lasers. Phys Lett A 53(1):77–78
8. Ohtsubo J (2012) Semiconductor lasers: stability, instability and chaos, vol 111. Springer
9. Van Tartwijk GH, Agrawal GP (1998) Laser instabilities: a modern perspective. Prog Quantum Electron 22(2):43–122
10. Bergé P, Pomeau Y, Vidal C (1986) Order in chaos: towards a deterministic approach to turbulence. Wile, New York
11. Arecchi F, Lippi G, Puccioni G, Tredicce J (1984) Deterministic chaos in laser with injected signal. Opts Commun 51(5):308–314
12. Weiss C, Abraham N, Hübner U (1988) Homoclinic and heteroclinic chaos in a single-mode laser. Phys Rev Lett 61(14):1587

13. Uchida A (2012) Optical communication with chaotic lasers: applications of nonlinear dynamics and synchronization. Wiley
14. Arecchi F, Meucci R, Puccioni G, Tredicce J (1982) Experimental evidence of subharmonic bifurcations, multistability, and turbulence in a Q-switched gas laser. Phys Rev Lett 49(17):1217
15. Dangoisse D, Glorieux P, Hennequin D (1987) Chaos in a CO_2 laser with modulated parameters: experiments and numerical simulations. Phys Rev A 36(10):4775
16. Paiella R, Martini R, Capasso F, Gmachl C, Hwang HY, Sivco DL, Baillargeon JN, Cho AY, Whittaker EA, Liu H (2001) High-frequency modulation without the relaxation oscillation resonance in quantum cascade lasers. Appl Phys Lett 79(16):2526–2528
17. Lorenz EN (1995) The essence of chaos. University of Washington Press
18. Epstein IR, Showalter K (1996) Nonlinear chemical dynamics: oscillations, patterns, and chaos. J Phys Chem 100(31):13 132–13 147
19. Becks L, Hilker FM, Malchow H, Jürgens K, Arndt H (2005) Experimental demonstration of chaos in a microbial food web. Nature 435(7046):1226
20. Nolte DD (2015) Introduction to modern dynamics: chaos, networks, space and time. Oxford University Press, USA
21. Vicente R, Daudén J, Colet P, Toral R (2005) Analysis and characterization of the hyperchaos generated by a semiconductor laser subject to a delayed feedback loop. IEEE J Quantum Electron 41(4):541–548
22. Wolf A (1986) Quantifying chaos with Lyapunov exponents. Chaos 16:285–317
23. Sano M, Sawada Y (1985) Measurement of the Lyapunov spectrum from a chaotic time series. Phys Rev Lett 55(10):1082
24. Wolf A, Swift JB, Swinney HL, Vastano JA (1985) Determining Lyapunov exponents from a time series. Physica D: Nonlinear Phenomena 16(3):285–317
25. Eckmann J-P, Kamphorst SO, Ruelle D, Ciliberto S (1986) Liapunov exponents from time series. Phys Rev A 34(6):4971
26. Bryant P, Brown R, Abarbanel HD (1990) Lyapunov exponents from observed time series. Phys Rev Lett 65(13):1523
27. Rosenstein MT, Collins JJ, De Luca CJ (1993) A practical method for calculating largest Lyapunov exponents from small data sets. Physica D: Nonlinear Phenomena 65(1–2):117–134
28. Grassberger P, Procaccia I (1983) Estimation of the Kolmogorov entropy from a chaotic signal. Phys Rev A 28(4):2591
29. Kaplan, JL Yorke JA (1979) Chaotic behavior of multidimensional difference equations. In: Functional differential equations and approximation of fixed points. Springer, pp 204–227
30. Pesin YB (1977) Characteristic Lyapunov exponents and smooth ergodic theory. Uspekhi Matematicheskikh Nauk 32(4):55–112
31. Wu J, Huang S-W, Huang Y, Zhou H, Yang J, Liu J-M, Yu M, Lo G, Kwong D-L, Duan S et al (2017) Mesoscopic chaos mediated by drude electron-hole plasma in silicon optomechanical oscillators. Nat Commun 8:15570
32. Grassberger P, Procaccia I (1983) Characterization of strange attractors. Phys Rev Lett 50(5):346
33. Lang R, Kobayashi K (1980) External optical feedback effects on semiconductor injection laser properties. IEEE J Quantum Electron 16(3):347–355
34. Grillot F (2009) On the effects of an antireflection coating impairment on the sensitivity to optical feedback of AR/HR semiconductor DFB lasers. IEEE J Quantum Electron 45(6):720–729
35. Olesen H, Osmundsen J, Tromborg B (1986) Nonlinear dynamics and spectral behavior for an external cavity laser. IEEE J Quantum Electron 22(6):762–773
36. Mork J, Semkow M, Tromborg B (1990) Measurement and theory of mode hopping in external cavity lasers. Electron Lett 26(9):609–610
37. Kakiuchida H, Ohtsubo J (1994) Characteristics of a semiconductor laser with external feedback. IEEE J Quantum Electron 30(9):2087–2097

38. Alsing PM, Kovanis V, Gavrielides A, Erneux T (1996) Lang and Kobayashi phase equation. Phys Rev A 53(6):4429
39. Erneux T, Kovanis V, Gavrielides A (2013) Nonlinear dynamics of an injected quantum cascade laser. Phys Rev E 88(3):032907
40. Liu T, Lee KE, Wang QJ (2013) Importance of the microscopic effects on the linewidth enhancement factor of quantum cascade lasers. Opt Express 21(23):27 804–27 815 (2013)
41. Hangauer A, Wysocki G (2015) Gain compression and linewidth enhancement factor in mid-IR quantum cascade lasers. IEEE J Sel Top Quantum Electron 21(6):74–84
42. Aellen T, Maulini R, Terazzi R, Hoyler N, Giovannini M, Faist J, Blaser S, Hvozdara L (2006) Direct measurement of the linewidth enhancement factor by optical heterodyning of an amplitude-modulated quantum cascade laser. Appl Phys Lett 89(9):091121
43. Glas P, Müller R, Klehr A (1983) Bistability, self-sustained oscillations, and irregular operation of a GaAs laser coupled to an external resonator. Opt Commun 47(4):297–301
44. Lenstra D, Verbeek B, Den Boef A (1985) Coherence collapse in single-mode semiconductor lasers due to optical feedback. IEEE J Quantum Electron 21(6):674–679
45. Kikuchi K, Okoshi T (1982) Simple formula giving spectrum-narrowing ratio of semiconductor-laser output obtained by optical feedback. Electron Lett 18(1):10–12
46. Agrawal G (1984) Line narrowing in a single-mode injection laser due to external optical feedback. IEEE J Quantum Electron 20(5):468–471
47. Lin C, Burrus C, Coldren L (1984) Characteristics of single-longitudinal-mode selection in short-coupled-cavity (SCC) injection lasers. J Light Technol 2(4):544–549
48. Gray GR, Huang D, Agrawal GP (1994) Chaotic dynamics of semiconductor lasers with phase-conjugate feedback. Phys Rev A 49(3):2096
49. Zah C, Osinski J, Menocal S, Tabatabaie N, Lee T, Dentai A, Burrus C (1987) Wide-bandwidth and high-power 1.3 μm InGaAsP buried crescent lasers with semi-insulating Fe-doped InP current blocking layers. Electron Lett 23(1):52–53
50. Jumpertz L (2017) Nonlinear photonics in Mid-infrared quantum cascade lasers. Springer
51. Tkach R, Chraplyvy A (1986) Regimes of feedback effects in 1.5-μm distributed feedback lasers. J Light Technol 4(11):1655–1661
52. Tkach R, Chraplyvy A (1985) Linewidth broadening and mode splitting due to weak feedback in single-frequency 1.5 μm lasers. Electron Lett 21(23):1081–1083
53. Fleming M, Mooradian A (1981) Spectral characteristics of external-cavity controlled semiconductor lasers. IEEE J Quantum Electron 17(1):44–59
54. Jumpertz L, Carras M, Schires K, Grillot F (2014) Regimes of external optical feedback in 5.6 μm distributed feedback mid-infrared quantum cascade lasers. Appl Phys Lett 105(13):131112
55. Kane DM, Shore KA (2005) Unlocking dynamical diversity: optical feedback effects on semiconductor lasers. Wiley
56. Donati S, Horng R-H (2012) The diagram of feedback regimes revisited. IEEE J Sel Top Quantum Electron 19(4):1 500 309–1 500 309
57. Jones R, Spencer P, Lawrence J, Kane D (2001) Influence of external cavity length on the coherence collapse regime in laser diodes subject to optical feedback. In: IEEE proceedings-optoelectronics, vol 148, no 1, pp 7–12
58. Sacher J, Elsässer W, Göbel EO (1989) Intermittency in the coherence collapse of a semiconductor laser with external feedback. Phys Rev Lett 63(20):2224
59. Heil T, Fischer I, Elsäßer W (1998) Coexistence of low-frequency fluctuations and stable emission on a single high-gain mode in semiconductor lasers with external optical feedback. Phys Rev A 58(4):R2672
60. Pan M-W, Shi B-P, Gray GR (1997) Semiconductor laser dynamics subject to strong optical feedback. Opt Lett 22(3):166–168
61. Mørk J, Mark J, Tromborg B (1990) Route to chaos and competition between relaxation oscillations for a semiconductor laser with optical feedback. Phys Rev Lett 65(16):1999
62. Fischer I, Van Tartwijk G, Levine A, Elsässer W, Göbel E, Lenstra D (1996) Fast pulsing and chaotic itinerancy with a drift in the coherence collapse of semiconductor lasers. Phys Rev Lett 76(2):220

63. Henry C, Kazarinov R (1986) Instability of semiconductor lasers due to optical feedback from distant reflectors. IEEE J Quantum Electron 22(2):294–301
64. Tromborg B, Mørk J, Velichansky V (1997) On mode coupling and low-frequency fluctuations in external-cavity laser diodes. Quantum Semiclassical Opt: J Eur Opt Soc Part B 9(5):831
65. Eguia MC, Mindlin GB, Giudici M (1998) Low-frequency fluctuations in semiconductor lasers with optical feedback are induced with noise. Phys Rev E 58(2):2636
66. Mørk J, Sabbatier H, Sørensen MP, Tromborg B (1999) Return-map for low-frequency fluctuations in semiconductor lasers with optical feedback. Opt Commun 171(1–3):93–97
67. Hegarty S, Huyet G, Porta P, McInerney J (1998) Analysis of the fast recovery dynamics of a semiconductor laser with feedback in the low-frequency fluctuation regime. Opt Lett 23(15):1206–1208
68. Van Tartwijk G, Levine A, Lenstra D (1995) Sisyphus effect in semiconductor lasers with optical feedback. IEEE J Sel Top Quantum Electron 1(2):466–472
69. Sano T (1994) Antimode dynamics and chaotic itinerancy in the coherence collapse of semiconductor lasers with optical feedback. Phys Rev A 50(3):2719
70. Seo D-S, Park J-D, McInerney JG, Osinski M (1989) Multiple feedback effects in asymmetric external cavity semiconductor lasers. IEEE J Quantum Electron 25(11):2229–2238
71. Hohl A, Van der Linden H, Roy R (1995) Determinism and stochasticity of power-dropout events in semiconductor lasers with optical feedback. Opt Lett 20(23):2396–2398
72. Torcini A, Barland S, Giacomelli G, Marin F (2006) Low-frequency fluctuations in vertical cavity lasers: experiments versus Lang-Kobayashi dynamics. Phys Rev A 74(6):063801
73. Aragoneses A, Rubido N, Tiana-Alsina J, Torrent M, Masoller C (2013) Distinguishing signatures of determinism and stochasticity in spiking complex systems. Sci Rep 3:1778
74. Jumpertz L, Schires K, Carras M, Sciamanna M, Grillot F (2016) Chaotic light at mid-infrared wavelength. Light: Sci Appl 5(6):e16088
75. Schwarz B (2010) Mapping the world in 3D. Nat Photon 4(7):429–430
76. Lin F-Y, Liu J-M (2004) Chaotic Lidar. IEEE J Sel Top Quantum Electron 10(5):991–997
77. Lin F-Y, Liu J-M (2004) Chaotic radar using nonlinear laser dynamics. IEEE J Quantum Electron 40(6):815–820
78. Cheng C-H, Chen C-Y, Chen J-D, Pan D-K, Ting K-T, F-Y Lin (2018) 3D pulsed chaos lidar system. Opt Express 26(9):12 230–12 241
79. Sciamanna M, Shore KA (2015) Physics and applications of laser diode chaos. Nat photon 9(3):151
80. Alspector J, Gupta B, Allen RB (1989) Performance of a stochastic learning microchip. In: Advances in neural information processing systems, pp 748–760
81. Marsaglia G, Zaman A (1993) Monkey tests for random number generators. Comput Math Appl 26(9):1–10
82. Schindler W, Killmann W (2002) Evaluation criteria for true (physical) random number generators used in cryptographic applications. In: International workshop on cryptographic hardware and embedded systems. Springer, pp 431–449
83. Oliver N, Soriano MC, Sukow DW, Fischer I (2013) Fast random bit generation using a chaotic laser: approaching the information theoretic limit. IEEE J Quantum Electron 49(11):910–918
84. Kanter I, Aviad Y, Reidler I, Cohen E, Rosenbluh M (2010) An optical ultrafast random bit generator. Nat Photon 4(1):58
85. Zhou Q, Valivarthi R, John C, Tittel W (2019) Practical quantum random-number generation based on sampling vacuum fluctuations. Quantum Eng 1(1):e8
86. Haw J-Y, Assad S, Lance A, Ng N, Sharma V, Lam PK, Symul T (2015) Maximization of extractable randomness in a quantum random-number generator. Phys Revw Appl 3(5):054004
87. Ma X, Yuan X, Cao Z, Qi B, Zhang Z (2016) Quantum random number generation. NPJ Quantum Inf 2:16021
88. Uchida A, Amano K, Inoue M, Hirano K, Naito S, Someya H, Oowada I, Kurashige T, Shiki M, Yoshimori S et al (2008) Fast physical random bit generation with chaotic semiconductor lasers. Nat Photon 2(12):728

89. Li N, Kim B, Chizhevsky V, Locquet A, Bloch M, Citrin D, Pan W (2014) Two approaches for ultrafast random bit generation based on the chaotic dynamics of a semiconductor laser. Opt Express 22(6):6634–6646
90. Tang X, Wu Z-M, Wu J-G, Deng T, Chen J-J, Fan L, Zhong Z-Q, Xia G-Q (2015) Tbits/s physical random bit generation based on mutually coupled semiconductor laser chaotic entropy source. Opt Express 23(26):33 130–33 141
91. Mezzapesa F, Columbo L, Brambilla M, Dabbicco M, Borri S, Vitiello M, Beere H, Ritchie D, Scamarcio G (2013) Intrinsic stability of quantum cascade lasers against optical feedback. Opt Express 21(11):13 748–13 757
92. Cardilli MC, Dabbicco M, Mezzapesa FP, Scamarcio G (2016) Linewidth measurement of mid infrared quantum cascade laser by optical feedback interferometry. Appl Phys Lett 108(3):031105
93. Shtengel G, Ackerman D, Morton P (1995) True carrier lifetime measurements of semiconductor lasers. Electron Lett 31(20):1747–1748
94. Trinité V, Ouerghemmi E, Guériaux V, Carras M, Nedelcu A, Costard E, Nagle J (2011) Modelling of electronic transport in quantum well infrared photodetectors. Infrared Phys Technol 54(3):204–208
95. Talukder MA, Menyuk CR (2011) Temperature-dependent coherent carrier transport in quantum cascade lasers. New J Phys 13(8):083027
96. Liu Z, Gmachl CF, Cheng L, Choa F-S, Towner FJ, Wang X, Fan J (2008) Temperature dependence of optical gain and loss in $\lambda \approx 8.2$–10.2 μm quantum-cascade lasers. IEEE J Quantum Electron 44(5):485–492
97. Carroll O, O'Driscoll I, Hegarty SP, Huyet G, Houlihan J, Viktorov EA, Mandel P (2006) Feedback induced instabilities in a quantum dot semiconductor laser. Opt Express 14(22):10 831–10 837
98. Viktorov E, Erneux T (2014) Self-sustained pulsations in a quantum-dot laser. Phys Rev E 90(5):052914
99. Preda CE, Ségard B, Glorieux P (2006) Weak temporal ratchet effect by asymmetric modulation of a laser. Opt Lett 31(15):2347–2349
100. Zamora-Munt J, Masoller C (2008) Generation of optical pulses in VCSELs below the static threshold using asymmetric current modulation. Opt Express 16(22):17 848–17 853
101. Watanabe N, Karaki K (1995) Inducing periodic oscillations from chaotic oscillations of a compound-cavity laser diode with sinusoidally modulated drive. Opt Lett 20(9):1032–1034
102. Buldú JM, García-Ojalvo J, Mirasso CR, Torrent M (2002) Stochastic entrainment of optical power dropouts. Phys Rev E 66(2):021106
103. Buldú J, Garcia-Ojalvo J, Torrent M (2004) Delay-induced resonances in an optical system with feedback. Phys Rev E 69(4):046207
104. Buldú JM, García-Ojalvo J, Torrent M, Vicente R, Pérez T, Mirasso CR (2003) Entrainment of optical low-frequency fluctuations is enhanced by coupling. Fluct Noise Lett 3(02):L127–L136
105. Toomey J, Kane DM, Lee M, Shore K (2010) Nonlinear dynamics of semiconductor lasers with feedback and modulation. Opt Express 18(16):16 955–16 972
106. Lawrence JS, Kane DM (2002) Nonlinear dynamics of a laser diode with optical feedback systems subject to modulation. IEEE J Quantum Electron 38(2):185–192
107. Karsaklian Dal Bosco A, Wolfersberger D, Sciamanna M (2013) Delay-induced deterministic resonance of chaotic dynamics. EPL (Europhysics Letters) 101(2):24001
108. Lam W-S, Guzdar PN, Roy R (2003) Effect of spontaneous emission noise and modulation on semiconductor lasers near threshold with optical feedback. Int J Mod Phys B 17(22, 24), 4123–4138
109. Aragoneses A, Sorrentino T, Perrone S, Gauthier DJ, Torrent M, Masoller C (2014) Experimental and numerical study of the symbolic dynamics of a modulated external-cavity semiconductor laser. Opt Express 22(4):4705–4713
110. Tiana-Alsina J, Quintero-Quiroz C, Panozzo M, Torrent M, Masoller C (2018) Experimental study of modulation waveforms for entraining the spikes emitted by a semiconductor laser with optical feedback. Opt Express 26(7):9298–9309

111. Pikovsky A, Kurths J, Rosenblum M, Kurths J (2003) In: Synchronization: a universal concept in nonlinear sciences, vol 12. Cambridge University Press
112. Longtin A (1997) Autonomous stochastic resonance in bursting neurons. Phys Rev E 55(1):868
113. Izhikevich EM (2000) Neural excitability, spiking and bursting. Int J Bifurc Chaos 10(06):1171–1266
114. Mendez JM, Laje R, Giudici M, Aliaga J, Mindlin G (2001) Dynamics of periodically forced semiconductor laser with optical feedback. Phys Rev E 63(6):066218
115. Sorrentino T, Quintero-Quiroz C, Aragoneses A, Torrent M, Masoller C (2015) Effects of periodic forcing on the temporally correlated spikes of a semiconductor laser with feedback. Opt Express 23(5):5571–5581
116. Sukow DW, Gauthier DJ (2000) Entraining power-dropout events in an external-cavity semiconductor laser using weak modulation of the injection current. IEEE J Quantum Electron 36(2):175–183
117. Rubido N, Tiana-Alsina J, Torrent M, Garcia-Ojalvo J, Masoller C (2011) Language organization and temporal correlations in the spiking activity of an excitable laser: experiments and model comparison. Phys Rev E 84(2):026202
118. Colet M, Aragoneses A (2018) Forecasting events in the complex dynamics of a semiconductor laser with optical feedback. Sci Rep 8(1):10741
119. Terazzi R, Gresch T, Giovannini M, Hoyler N, Sekine N, Faist J (2007) Bloch gain in quantum cascade lasers. Nat Phys 3(5):329
120. Takiguchi Y, Liu Y, Ohtsubo J (1998) Low-frequency fluctuation induced by injection-current modulation in semiconductor lasers with optical feedback. Opt Lett 23(17):1369–1371
121. Buldú JM, Chialvo DR, Mirasso CR, Torrent M, García-Ojalvo J (2003) Ghost resonance in a semiconductor laser with optical feedback. EPL (Europhysics Letters) 64(2):178
122. Cronin-Golomb M, Lau KY, Yariv A (1985) Infrared photorefractive passive phase conjugation with $BaTiO_3$: demonstrations with GaAlAs and 1.09-μm Ar^+ lasers. Appl Phys Lett 47(6):567–569
123. Miltenyi E, Ziegler M, Hofmann M, Sacher J, Elsässer W, Göbel E, MacFarlane D (1995) Long-term stable mode locking of a visible diode laser with phase-conjugate feedback. Opt Lett 20(7):734–736
124. Kürz P, Mukai T (1996) Frequency stabilization of a semiconductor laser by external phase-conjugate feedback. Opt Lett 21(17):1369–1371
125. Vahala K, Kyuma K, Yariv A, Kwong S-K, Cronin-Golomb M, Lau KY (1986) Narrow linewidth, single frequency semiconductor laser with a phase conjugate external cavity mirror. Appl Phys Lett 49(23):1563–1565
126. Petersen L, Gliese U, Nielsen TN (1994) Phase noise reduction by self-phase locking in semiconductor lasers using phase conjugate feedback. IEEE J Quantum Electron 30(11):2526–2533
127. Agrawal GP, Klaus JT (1991) Effect of phase-conjugate feedback on semiconductor laser dynamics. Opt Lett 16(17):1325–1327
128. Yariv A (1978) Phase conjugate optics and real-time holography. IEEE J Quantum Electron 14(9):650–660
129. Purcell EM (1977) Life at low Reynolds number. Am J Phys 45(1):3–11
130. Fink M (1992) Time reversal of ultrasonic fields. I. Basic principles. IEEE Trans Ultrason Ferroelectr Freq Control 39(5):555–566
131. Virte M, Karsaklian Dal Bosco A, Wolfersberger D, Sciamanna M (2011) Chaos crisis and bistability of self-pulsing dynamics in a laser diode with phase-conjugate feedback. Phys Rev A 84(4):043836
132. Cronin-Golomb M, Fischer B, White JO, Yariv A (1984) Theory and applications of four-wave mixing in photorefractive media. IEEE J Quantum Electron 20(1):12–30
133. Krauskopf B, Gray GR, Lenstra D (1998) Semiconductor laser with phase-conjugate feedback: dynamics and bifurcations. Phys Rev E 58(6):7190

134. Pieroux D, Erneux T, Haegeman B, Engelborghs K, Roose D (2001) Bridges of periodic solutions and tori in semiconductor lasers subject to delay. Phys Rev Lett 87(19):193901
135. Mercier É, Wolfersberger D, Sciamanna M (2016) High-frequency chaotic dynamics enabled by optical phase-conjugation. Sci Rep 6:18988
136. Bouchez G, Uy C-H, Macias B, Wolfersberger D, Sciamanna M (2019) Wideband chaos from a laser diode with phase-conjugate feedback. Opt Lett 44(4):975–978
137. Rosencher E, Vinter B (2002) Optoelectronics. Cambridge University Press
138. Yang RQ, Xu J, Sweeny M (1994) Selection rules of intersubband transitions in conduction-band quantum wells. Phys Rev B 50(11):7474
139. Liu H, Buchanan M, Wasilewski Z (1998) How good is the polarization selection rule for intersubband transitions? Appl Phys Lett 72(14):1682–1684
140. Janassek P, Hartmann S, Molitor A, Michel F, Elsäßer W (2016) Investigations of the polarization behavior of quantum cascade lasers by Stokes parameters. Opt Lett 41(2):305–308
141. Van Exter M, van Doorn AJ, Woerdman J (1997) Electro-optic effect and birefringence in semiconductor vertical-cavity lasers. Phys Rev A 56(1):845
142. Valle A, Shore KA, Pesquera L (1996) Polarization selection in birefringent vertical-cavity surface emitting lasers. J Lightwave Technol 14(9):2062–2068
143. Nishiyama N, Mizutani A, Hatori N, Arai M, Koyama F, Iga K (1999) Lasing characteristics of InGaAs-GaAs polarization controlled vertical-cavity surface-emitting laser grown on GaAs [311] B substrate. IEEE J Sel Top Quantum Electron 5(3):530–536
144. Williams QL, Roy R (1996) Fast polarization dynamics of an erbium-doped fiber ring laser. Opt Lett 21(18):1478–1480
145. Nieubor N, Panajotov K, Goulet A, Veretennicoff I, Thienpont H (1998) Data transparent reconfigurable optical interconnections based on polarization-switching VCSELs and polarization-selective diffractive optical elements. IEEE Photonics Technol Lett 10(7):973–975
146. Katayama T, Ooi T, Kawaguchi H (2009) Experimental demonstration of multi-bit optical buffer memory using 1.55 μm polarization bistable vertical-cavity surface-emitting lasers. IEEE J Quantum Electron 45(11):1495–1504
147. Oliver N, Soriano MC, Sukow DW, Fischer I (2011) Dynamics of a semiconductor laser with polarization-rotated feedback and its utilization for random bit generation. Opt Lett 36(23):4632–4634
148. Masoller C, Torre M (2005) Influence of optical feedback on the polarization switching of vertical-cavity surface-emitting lasers. IEEE J Quantum Electron 41(4):483–489
149. Höpfner H, Lindemann M, Gerhardt NC, Hofmann MR (2014) Controlled switching of ultrafast circular polarization oscillations in spin-polarized vertical-cavity surface-emitting lasers. Appl Phys Lett 104(2):022409
150. Pérez P, Lin H, Valle Á, Pesquera L (2014) Polarization dynamics induced by orthogonal optical injection close to the lasing mode of a single-transverse-mode VCSEL. JOSA B 31(11):2901–2907
151. Gavrielides A, Erneux T, Sukow DW, Burner G, McLachlan T, Miller J, Amonette J (2006) Square-wave self-modulation in diode lasers with polarization-rotated optical feedback. Opt Lett 31(13)
152. Shibasaki N, Uchida A, Yoshimori S, Davis P (2006) Characteristics of chaos synchronization in semiconductor lasers subject to polarization-rotated optical feedback. IEEE J Quantum Electron 42(3):342–350
153. Takeuchi Y, Shogenji R, Ohtsubo J (2008) Chaotic dynamics in semiconductor lasers subjected to polarization-rotated optical feedback. Appl Phys Lett 93(18):181105
154. Sukow DW, Gavrielides A, Erneux T, Mooneyham B, Lee K, McKay J, Davis J (2010) Asymmetric square waves in mutually coupled semiconductor lasers with orthogonal optical injection. Phys Rev E 81(2):025206
155. Gavrielides A, Sukow DW, Burner G, McLachlan T, Miller J, Amonette J (2010) Simple and complex square waves in an edge-emitting diode laser with polarization-rotated optical feedback. Phys Rev E 81(5):056209

156. Uy C-H, Weicker L, Rontani D, Sciamanna M (2018) Sustained oscillations accompanying polarization switching in laser dynamics. Opt Express 26(13):16 917–16 924
157. Brosseau C (1998) Fundamentals of polarized light. Wiley, New York
158. Molitor A (2015) Investigations on Stokes polarization parameters of vertical-cavity surface-emitting lasers: spatially, spectrally and time-resolved. PhD dissertation, Technische Universität Darmstadt
159. Uy C-H (2018) Two-polarization dynamics in optically delayed lasers. PhD dissertation, CentraleSupélec
160. Bezerianos A, Bountis T, Papaioannou G, Polydoropoulos P (1995) Nonlinear time series analysis of electrocardiograms. Chaos: Interdiscip J Nonlinear Sci 5(1):95–101
161. Minati L, Chiesa P, Tabarelli D, D'Incerti L, Jovicich J (2015) Synchronization, non-linear dynamics and low-frequency fluctuations: analogy between spontaneous brain activity and networked single-transistor chaotic oscillators. Chaos: Interdisci J Nonlinear Sci 25(3):033107
162. Pacifici D, Lezec HJ, Atwater HA (2007) All-optical modulation by plasmonic excitation of CdSe quantum dots. Nat Photonics 1(7):402
163. Ríos C, Stegmaier M, Hosseini P, Wang D, Scherer T, Wright CD, Bhaskaran H, Pernice WH (2015) Integrated all-photonic non-volatile multi-level memory. Nat Photonics 9(11):725
164. Dillane M, Tykalewicz B, Goulding D, Garbin B, Barland S, Kelleher B (2019) Square wave excitability in quantum dot lasers under optical injection. Opt Lett 44(2):347–350

Chapter 4
Chaos Synchronization and Its Application to Secure Communications

Since the beginning of the 90ies, chaos has been viewed as a powerful tool for several applications. The complexity of the chaotic pattern is promising for the generation of true random numbers. Because of its unpredictability, chaos also shed the light on secure communications, especially in the optoelectronics field. Several methods of encryption were studied with a particular focus on the chaos being the carrier among which the message is hidden. This method relies on synchronization (or anti-synchronization) of chaos and implies one emitter and one receiver. This technique is well mastered when it comes to fiber optics, with the latest efforts achieving high data rates and low bit-error rates (BER). At the opposite, there are only a few theoretical reports for free-space chaos synchronization and communications, and experiments are never mentioned, though they could be very relevant.

The beginning of this chapter will introduce the techniques that were first envisioned for chaos communication. Then, I will describe the state-of-the-art chaos synchronization in laser diodes and how to use this synchronization for fibered secure communications. The QCLs chaos previously detailed will be used to extend the known properties of fibered secure communications to free-space secure communications. Further options will be discussed in order to improve the data rate of the encrypted transmission, that will focus on optical injection. A summary of the chaotic secure transmission process can be found in Fig. 4.1.

O. Spitz, *Mid-infrared Quantum Cascade Lasers for Chaos Secure Communications*, Springer Theses, https://doi.org/10.1007/978-3-030-74307-9_4

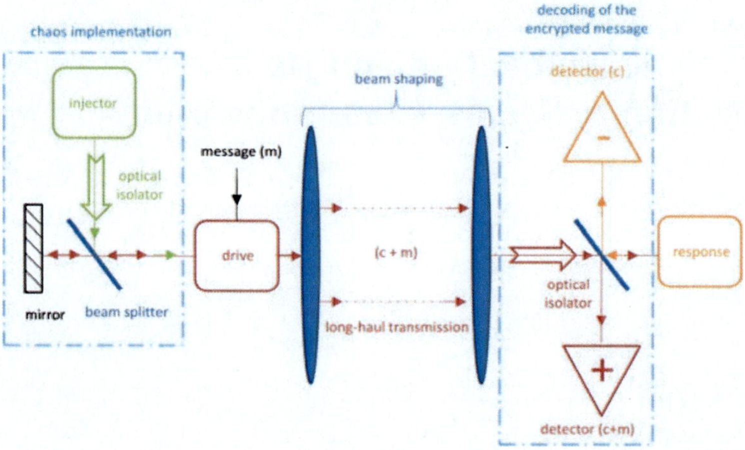

Fig. 4.1 Schematic of optical chaos synchronization and secure communication between a drive laser and a response laser. The drive is made chaotic with external feedback and/or injection. The chaotic light of the drive is then injected into the response through an optical isolator so that the light from the response does not interfere with the drive. Most of the experimental setups are more complex than this one because they include, for instance, optical fibers, EDFA amplifiers, etc. Details for each configurations can be found in the papers cited below

4.1 Taking Advantage of Chaos

4.1.1 Towards Experimental Synchronization

The first intuition of chaos application was unveiled in 1990 with the experimental synchronization of two chaotic electronic circuits [1] and followed many efforts carried out in the 80ies about the implementation of chaos in electronics circuits, such as the Chua's circuit [2, 3] detailed in Fig. 4.2. Pecora and Carroll showed that it was possible to have a temporal reproducible behavior between two electronic circuits even if there was a mismatch between the circuits, as shown in Fig. 4.3. Synchronization describes how a receiver is capable of following the dynamical properties of an emitter, when the emitter and the receiver are coupled, as illustrated in Fig. 4.4. The synchronization of periodic oscillations was first described by Christian Huygens in the seventeenth century. Two pendulum clocks sharing the same support had their oscillations coincided perfectly [4]. The concept has then been extended to several domains and for instance, optical injection-locking between two lasers can be viewed as synchronization of periodic optical-carrier frequency.

In a similar fashion to what was proposed to describe chaos, we restrain ourselves to a discrete model and explain the general idea of chaos synchronization with a simple differential system. The chaotic system under study is composed of vector variables $\vec{u}$ and $\vec{v}$. The transmitter system is divided into two subsystems and describe

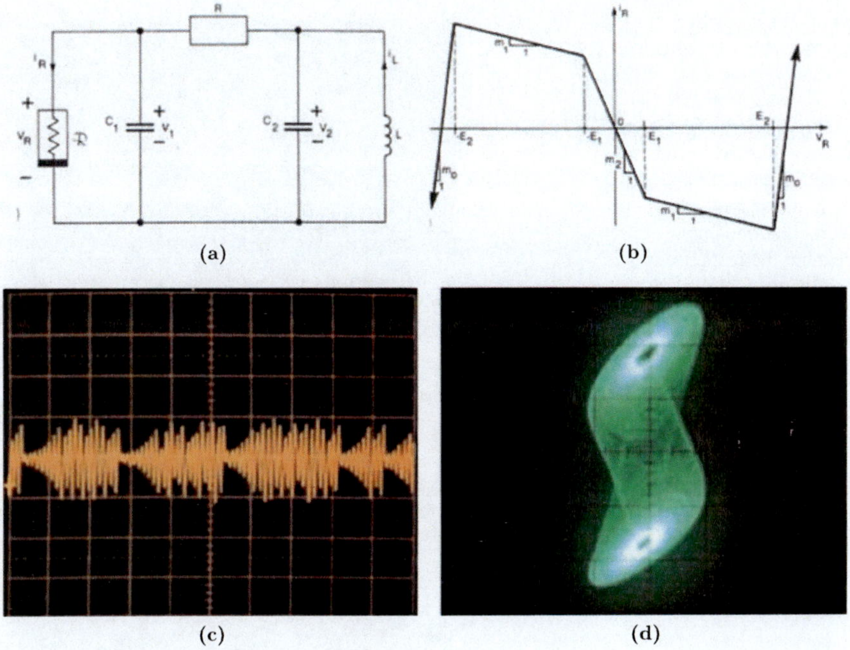

Fig. 4.2 Chua's circuit (a) and the 5-segment $v_{\Re} - i_{\Re}$ characteristic (b) for the non-linear resistor $\Re$. For computer simulations, chaos can be observed with only the 3 inner negative slope segments, as visualized in the characteristic. c Typical chaotic waveform that can be displayed with Chua's circuit and d the related phase-space diagram retrieved with an oscilloscope and showing two attractors. Courtesy of [2, 3]

the nonlinearities for the respective subsystems as $f(\vec{u}, \vec{v})$ and $g(\vec{u}, \vec{v})$. Then, the transmitter system is characterized with the set of equations:

$$\frac{d\vec{u}}{dt} = f(\vec{u}, \vec{v}) \tag{4.1}$$

$$\frac{d\vec{v}}{dt} = g(\vec{u}, \vec{v}) \tag{4.2}$$

The subsystems are mutually coupled and the outputs of $\vec{u}$ and $\vec{v}$ are assumed to be chaotic. We prepare the receiver system for chaos synchronization. The receiver system is composed of only one of the subsystems and it is written in a similar manner than the two previous equations. That is to say:

$$\frac{d\vec{w}}{dt} = f(\vec{w}, \vec{v}) \tag{4.3}$$

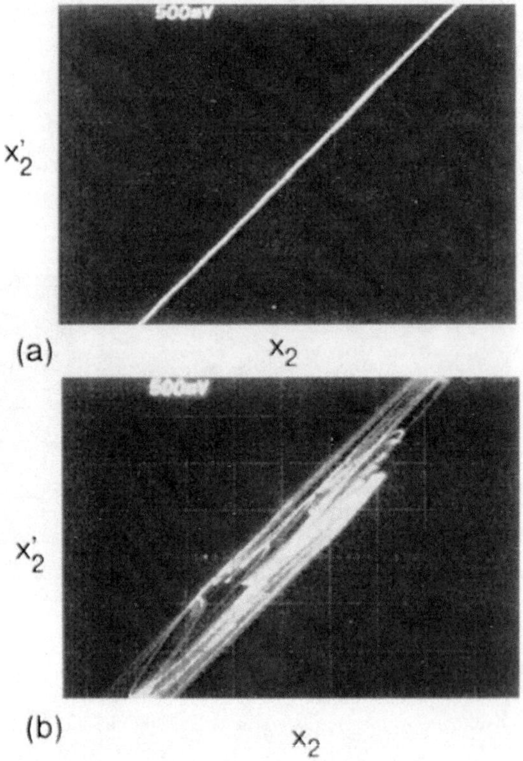

Fig. 4.3 **a** Correlation diagram in the case where the parameters of the response exactly match that of the drive and **b** correlation diagram when the circuit parameters are different by 50%. In the second case, synchronization is not perfect but the response is still influenced by the drive. Courtesy of [1]

Without coupling between drive and response, variable $\vec{v}$ is treated as a certain constant vector. Consequently, the output $\vec{w}$ may be either a chaotic or a stable oscillation when the response receives no transmitted signal. Even if the drive output is chaotic, the response output $\vec{w}$ would never show the same chaotic oscillations because chaos has strict sensitivity for initial conditions. However, the response output $\vec{w}$ exhibits the same chaotic oscillations as the drive output $\vec{u}$ as soon as a fraction of the drive output is sent to the receiver under appropriate conditions. From a mathematical point of view, there is a condition for having a negative value for the largest Lyapunov exponent concerning the difference in the outputs $\vec{u}$ and $\vec{w}$.

If we take for instance a Lorenz system similar to the one we used to describe chaos, the differential equations are expressed with the variables x, y and z and the drive system is divided into two parts, namely:

$$\begin{cases} \frac{dy}{dt} = -xz + rx - y \\ \frac{dz}{dt} = xy - bz \end{cases} \tag{4.4}$$

and

$$\frac{dx}{dt} = \sigma(y - x) \tag{4.5}$$

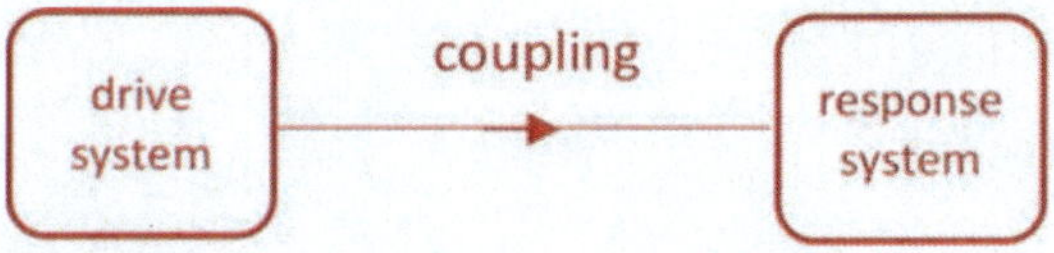

Fig. 4.4 Scheme illustrating the coupling between a drive system and a response system that leads to a synchronization of the response with the drive's properties

The response system evolves as the drive system, except that the variable x' is treated as a constant without the coupling from the drive:

$$\begin{cases} \frac{dy'}{dt} = -x'z' + rx' - y' \\ \frac{dz'}{dt} = x'y' - bz' \end{cases} \tag{4.6}$$

We focus on an example from Ref. [5] where the outputs x, y, and z of the drive are chaotic at the parameter values of $\sigma = 16$, $b = 4$ and $r = 45.92$. Because initially, the coupling is not effective, the maximum Lyapunov exponent of the response system is negative. Therefore, the response exhibits stable outputs without receiving any chaotic signal from the drive. Then, the response is coupled with the drive and we follow the evolution of the trajectories in the (y, z) plane and the (y', z') plane. Figure 4.5 shows that, starting from the coupling instant, the distance between the two trajectories is decreased and the trajectories eventually overlap each other. Chaos synchronization is therefore achieved. It is possible to reach chaos synchronization only for a limited set of parameters. One of the key elements is that the response system must be very similar to the drive system when one or more of the drive outputs are sent to the response.

In laser systems, however, the method is not straightforwardly applicable, since it is not possible to divide the dynamics of laser variables into subsystems. What is actually done is that two similar laser systems are prepared as drive and response, and a small portion of the drive output is injected into the response. Even if the synchronization of periodic waveforms can be clearly understood with the example of the coupled pendula, it is more difficult to apprehend the synchronization of chaotic waveforms. Indeed, chaos is very sensitive to initial conditions and a very small deviation between two chaotic systems leads to exponentially divergent behaviors. In the case of experimental optoelectronics where all the parameters cannot be exactly reproduced from one laser to another, it thus seems nearly impossible to obtain a synchronization in which the drive and the response systems exactly behave the same. Against all odds, it is possible to synchronize their chaotic dynamics starting from arbitrary initial conditions, provided the two setups are sufficiently similar and the coupling between the two lasers is strong enough. If there is a parameter mismatch, the quality of the synchronization degrades until it is not possible to link the response output with the drive output. This conclusion about the possibility of synchronization of optical chaos was pushed on by the first numerical study of that kind [6], also in 1990. Winful and Rahman showed that an array of injection-coupled semiconductor lasers were able to have the same chaotic output. These efforts paved the way for the

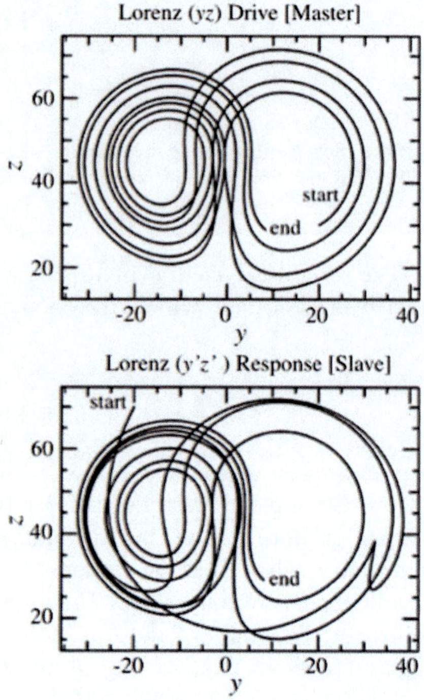

Fig. 4.5 Schematic of chaos synchronization in the case where the initial conditions for the response and the drive are different. Nevertheless, the response follows the chaotic path of the drive because the coupling between the two is strong. Courtesy of [5]

first experimental proof of chaos synchronization in an optical system. In 1994, three different groups showed an experimental chaos synchronization with three different setups. Roy and Thornburg used mutually coupled Nd:YAG solid-state lasers [7]. One or both lasers are driven chaotic by periodic modulation of their pump laser and the synchronization occurs when the lasers are strongly coupled. A Japanese team carried out a similar experiment with unidirectionally coupled CO_2 lasers [8]. The chaotic behavior observed in the drive and the response laser is caused by a saturable absorber inducing self-sustained pulsations. Synchronization occurs when the output of the drive laser is unidirectionally injected into the saturable absorber of the response laser, provided that the intensity of the drive is strong enough. With a very similar setup, Liu and co-workers showed that the quality of synchronization in two gas lasers with an intracavity saturable absorber was largely dependent on the optical frequency detuning between the drive laser (also called master laser) and the response laser (also called slave laser) [9]. One of their optical synchronization results can be found in Fig. 4.6. It is indeed relevant to note that in optical chaos, there are two types of frequencies involved in the electric field of the laser output. One of these two frequencies is related to the wavelength of the optical emission. As already mentioned, in the case of mid-infrared QCLs, this frequency ranges between 30 and 75 THz because the wavelength ranges between 4 and 10 μm. This fast frequency can be considered as the carrier and the dynamics frequency observed in the waveforms can be considered as the envelope. In QCLs, this slow frequency ranges between

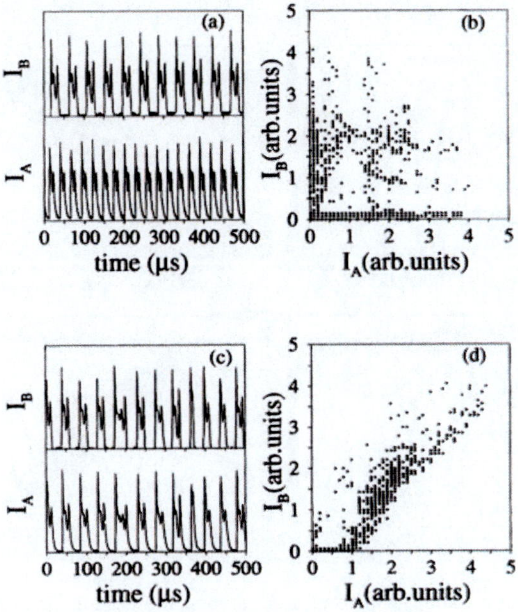

Fig. 4.6 One of the pioneering experimental demonstrations of chaos synchronization in an optical system. **a** Before coupling, the two lasers are pulsing but there is no sign of synchronization in the correlation diagram. **b** Intensities' correlation of uncoupled lasers. After coupling **c** the chaotic lasers become synchronized and their intensities show **d** a positive degree of correlation. Courtesy of [9]

0.1 and 100 MHz, as we saw in the subsection dealing with the LFF phenomenon, and in other semiconductor lasers, this frequency rarely exceeds 10 GHz, mainly because of relaxation frequency concerns. Even if it is only possible to observe the slow frequency with real-time detectors (MCT detectors or QWIPs in the case of QCLs), both the slow frequency and the fast frequency must be synchronized. The first one is usually observed in the timetrace and the related synchronization can be easily assessed with a correlation diagram. The correlation diagram represents the output power of the slave laser as function of the master laser output. The second one can be evaluated with a FTIR and, when synchronization occurs, the optical frequencies match due to frequency-pulling effect [10]. This is called injection-locking between two coupled lasers. The synchronization of the slow frequency can only be possible if the master and the slave laser are in the injection-locking range. After being analyzed in the mechanical, electrical and optical domains, the concept of chaos synchronization was extended to the biological domain with the study of coupled bursting neurons [11, 12]. This is another example of the interactions between optical networks and biological clusters and this gives credit to the many efforts aiming at reproducing a system as intricate as the brain with the help of complex photonics [13].

It is relevant to note that chaos anti-synchronization can also occur when injecting the chaotic fluctuations of the drive into the response laser, as already shown in

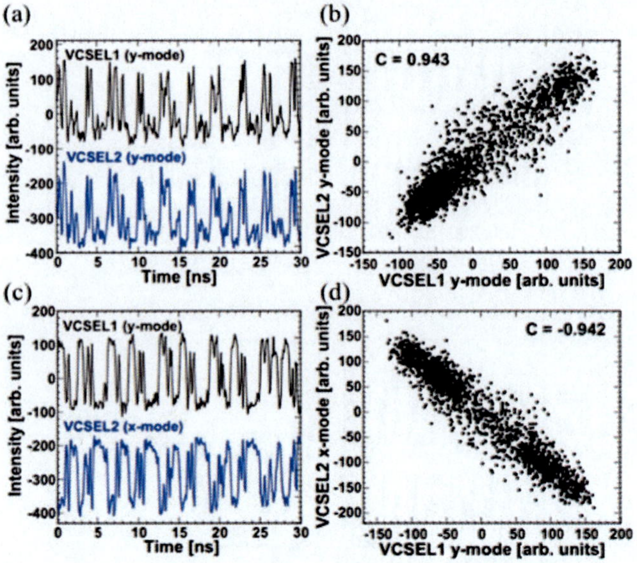

Fig. 4.7 Chaos synchronization (top row) and anti-synchronization (bottom row) in coupled VCSELs. The left column shows the photon intensities for the two polarizations of the laser and the right column shows the correlation diagram. In the case of synchronization, the correlation diagram shows a positive correlation and in the case of anti-synchronization, the correlation diagram shows a negative correlation. Courtesy of [14]

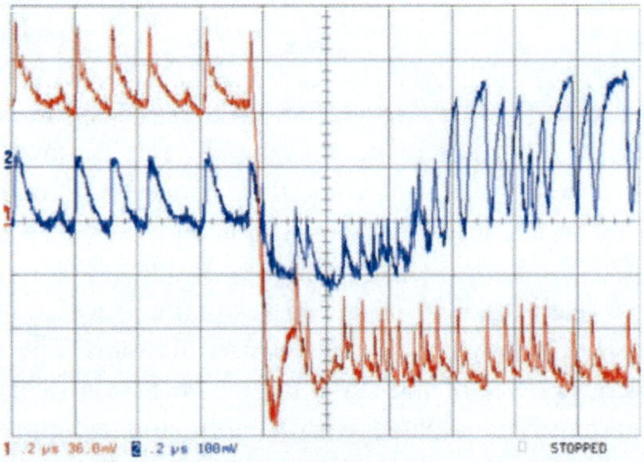

Fig. 4.8 Switching between synchronization and anti-synchronization by changing the pump current of the master laser. The red signal is that of the master laser and the blue signal is that of the slave laser. This emphasizes that the boundaries between synchronization and anti-synchronization can be thin. Courtesy of [15]

VCSELs and laser diodes [14, 15] and illustrated in Fig. 4.7. Switching between both types of synchronization is possible, for example, by changing slightly the pump current of the drive system, as visualized in Fig. 4.8. In this way, such a configuration can be used for chaos communications without synchronization, as explained below, in addition to the methods related to the entrainment phenomenon and the on-demand extreme events which can also be useful to control spiking dynamics. Mathematically, it is possible to give more details about different kinds of synchronizations which can be found in coupled lasers. Several configurations such as complete synchronization [16], generalized synchronization [17], anticipating synchronization [18] and phase synchronization [19] exist.

4.1.2 Towards Secure Communications

Even if chaos synchronization is a mind-blowing concept, it has no direct application. However, the earliest papers about chaos synchronization already envisioned a glimpse of the potential of this phenomenon, especially for secure communications [20]. The main idea was to hide a message with a small amplitude within the chaotic signal of the transmitter so that the message does not disturb the larger chaotic fluctuations and remains well concealed from an eavesdropper. Indeed, synchronization is robust to small changes in the coupling strength and consequently, the response laser will still synchronize to the chaos of the drive laser, even in the presence of a perturbation that is due to message encoding. This mixed signal is then sent to the receiver and a detector. As shown previously, in the case of optical chaos, the receiver must be very similar to the transmitter so that the synchronization occurs with a correlation diagram looking like a straight line at 45°. As the receiver synchronizes only with the chaotic fluctuations, subtracting the signal of the receiver from that of the detector allows recovering the hidden message. The first demonstration of such transmission in electronics circuit is presented in Fig. 4.9 and showed it was

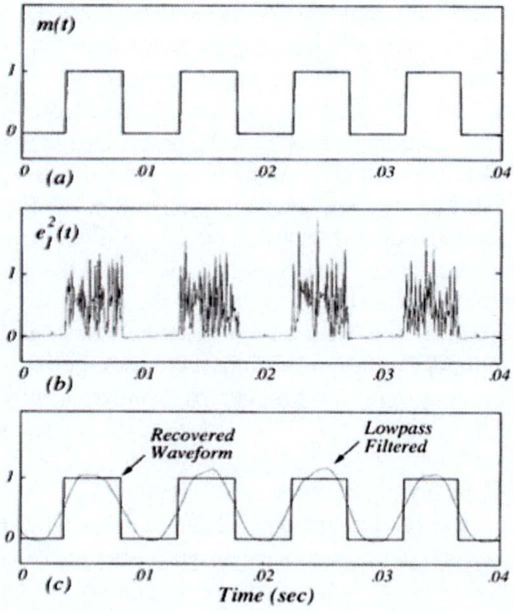

Fig. 4.9 First example of secure communication achieved through chaos synchronization. **a** is the message to be transmitted, **b** is the recovered signal after the subtraction and **c** is the deciphered message after low-pass filtering. Courtesy of [21]

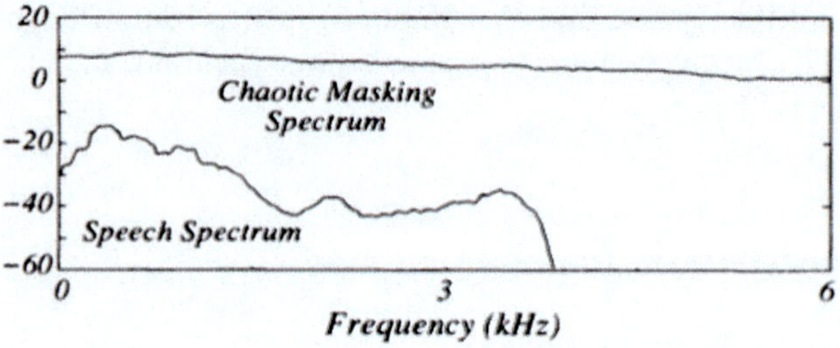

Fig. 4.10 In order to properly encipher the message within the chaotic fluctuations, the amplitude of the message must be small compared to the amplitude of the chaos. The same argument is true for the maximum frequency of the message, otherwise it can be detected. The speech spectrum to be concealed meets these requirements regarding the properties of the chaotic signal. Courtesy of [21]

possible to recover a speech message, in 1993 [21]. This experimental effort also exposed that the message to be transmitted must be small in terms of amplitude and frequency bandwidth compared to the masking chaos, as shown in Fig. 4.10. The idea was extended to the optical domain one year later, with the first numerical proposal of optical secure communication [22]. In a similar scheme, a chaotic spiking solid-state laser was used as the drive and the message was recovered through synchronization and subtraction. These first numerical studies achieved a secure data transmission at a few kbits/s and there was a growing demand for transmission at higher bit-rates. This was compatible with the use of optical chaos because the latter had already exhibited bandwidth in the order of a few hundreds of MHz [23, 24], which paved the way for secure transmissions at several dozens of Mbits/s. In 1996, two numerical models with realistic operational characteristics showed a secure transmission with synchronized chaotic semiconductor lasers at a data rate of 5 Mbits/s [25] and 300 Mbits/s [26]. As these simulations were carried out with near-infrared semiconductor lasers, the transmission occurs through a fiber and this means that long-distance secure communications can be envisioned. Mirasso and co-workers implemented a 50 km fiber transmission with amplification in order to compensate the losses in the fiber. This numerical analysis, as well as the following efforts from the same group (see for instance Fig. 4.11), paved the way for the first experimental proof of optical secure communication, because it proved the high potential of synchronized chaotic semiconductor lasers for long-range high-speed encrypted data transmissions. This demonstration occurred two years later and, as always in the history of chaos secure communications, two different groups reported an experimental secure transmission the same year. The first one transmitted a message at 10 MHz with erbium-doped fiber-ring lasers [27]. As it has been explained numerically, the message hidden within the chaotic carrier could be retrieved because the message is small enough and does not mainly perturb the chaotic fluctuations. The master laser thus synchronizes with the slave laser on the one hand, and a detector retrieves the signal of the master laser with the message concealed within the chaos on the other hand. A subtraction of the detector's signal and slave's signal allows recovering the message. A filtering is often needed in order to improve the quality of the transmission. The second one took advantage of the chaotic fluctuations generated by a semiconductor laser with an optoelectronic feedback loop [28]. The message to be transmitted is introduced in the feedback loop which also provides the time-delayed signal driving the master laser chaotic. The slave laser (which is not fed back) receives the signal from the master laser and synchronizes with the chaotic signal only. The first experiment with semiconductor lasers was unveiled in 1999. Sivaprakasam and Shore demonstrated a secure communication with diode lasers under EOF but the data rate was limited to a few kbits/s [29]. The maximum speed transmission was then experimentally increased in chaotic diode lasers to a few Gbits/s corresponding to a limit set by the relaxation frequency for this kind of lasers [30]. Even if the possibility of high-speed secure transmissions was demonstrated, the question about the evolution of the BER in a long-scale scheme remained open. Simulations had shown a degradation of the performances, such as the correlation diagram and the eye diagram, when increasing the fiber length from 50 to 200 km [31], as shown in Fig. 4.12. In the real-world conditions of the fiber-optics network of Athens, the performance of all-optical fiber secure communications systems has been investigated with chaotic semiconductor lasers under EOF [32]. Through the whole length of the 120 km fiber system, Argyris and co-workers were able to achieve a secure transmission at 2.4 Gbits/s with a BER of 10^{-3} and at 1 Gbits/s with a BER of 10^{-7}. Figure 4.13 gathers the BER performances of this field experiment. These findings allowed the creation of a company called Eulambia and dedicated to the implementation of fibered secure communications and RNG. Simultaneously, an Italian group experimented a composite video signal transmission over 1.2 km [33]. The carrier frequency of the signal is at 2.4 GHz and a compromise has to be found between a sufficient masking of the signal within the chaotic carrier and a decent video quality after recovery and subtraction, as shown in Fig. 4.14. Consequently, there is a 4-dB discrepancy between the signal of the enciphered message and that of the chaotic carrier.

Another key parameter is the technique used to encipher the message within the chaotic carrier. Three methods have been currently explored and they have a tremendous influence on the BER that can be achieved with a given laser configuration [34], as can be seen in Fig. 4.15. An overview of the performances of the three methods can also be found in Fig. 4.16.

The first method is called chaos masking (CMa). It consists in embedding a small message $m(t)$ into a chaotic carrier $x(t)$ at the drive level and then, sending the signal $x(t) + m(t)$ to the response. The response system share similarities with the drive and the two systems are operating at the same parameter values. Only the chaotic signal $x(t)$ is reproduced in the response system if the amplitude of the message is small enough. Then, the message $m(t)$ is deciphered by subtracting the response output $x(t)$ from the transmission signal $x(t) + m(t)$. The success of the message secure concealing into a chaotic carrier and of the decoding of the message relies on a message amplitude sufficiently small compared with the averaged chaotic carrier signal. Usually, the fraction is less than 1% of the average chaotic power.

The second method is called chaos modulation (CMo) and was introduced in 1993 [35]. In this method, a message is inserted within a chaotic carrier in the non-linear oscillator and the two signals conform a new chaotic state different from the original

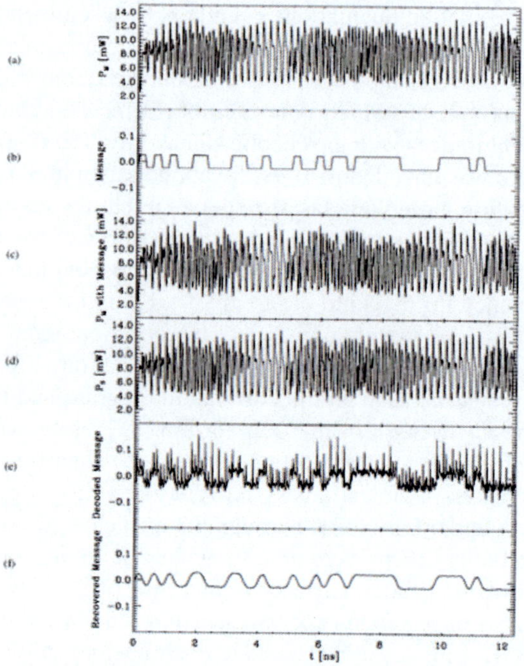

Fig. 4.11 Simulation of the dynamical evolution of the master diode laser and the slave diode laser linked by a fiber of 50 km. **a** chaotic output from the master laser. **b** message to be transmitted at a bit-rate of 4 Gbit/s. **c** Master laser output after the encoding of the message. **d** Output of the slave laser synchronized with the master laser. **e** Decoded message by subtraction. **f** Decoded message after subtraction and filtering. Courtesy of [31]

one. In CMo, a delayed feedback system is usually used as a chaotic generator. This new signal together with the message is sent to the response. Since the drive and the response are the same non-linear systems, the chaotic oscillation is exactly reproduced in the response system thanks to chaos synchronization. By subtracting the synchronized chaotic signal from the transmitted signal, it is possible to decode the message. Sometimes, the message is decoded by dividing the transmitted signal by the synchronized chaotic signal in the response. Contrary to the previous method, CMo has no restriction on the magnitude of the message as a secure communication, since both the chaotic carrier and the message conform new chaotic states in the non-linear systems. However, the degree of security for data transmissions becomes worse when the signal level of a message is large. Therefore, the amplitude of a message in CMo should also be small.

The last method is called chaos shift keying (CSK) and was also first demonstrated in 1993 [36]. The signal shift keying technique, which is frequently used in ordinary communication systems, is also applicable to chaotic data transmissions. In CSK, two chaotic states $x_1(t)$ and $x_2(t)$ are generated in a drive system. The switching itself to send either chaotic state is a message $m(t)$. In the response system, each state $x_1(t)$

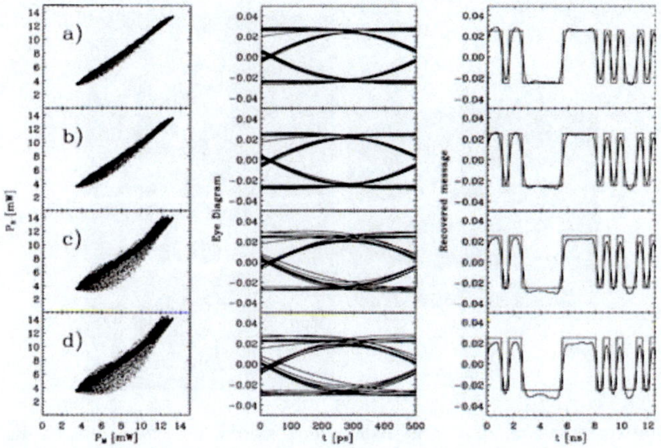

Fig. 4.12 Numerical analysis of a long-distance secure transmission through an optical fiber. Left column, correlation diagram. Center column, eye diagram of the recovered message after decoding and filtering. Right column, encoded and recovered message. The first row corresponds to a 50-km fiber transmission, the second row a 100-km fiber transmission, the third row a 150-km fiber transmission and the fourth row a 200-km fiber transmission. The message has been modulated at a bit-rate of 2 Gbit/s and the right column shows a degradation of the recovery process when increasing the fiber length. Courtesy of [31]

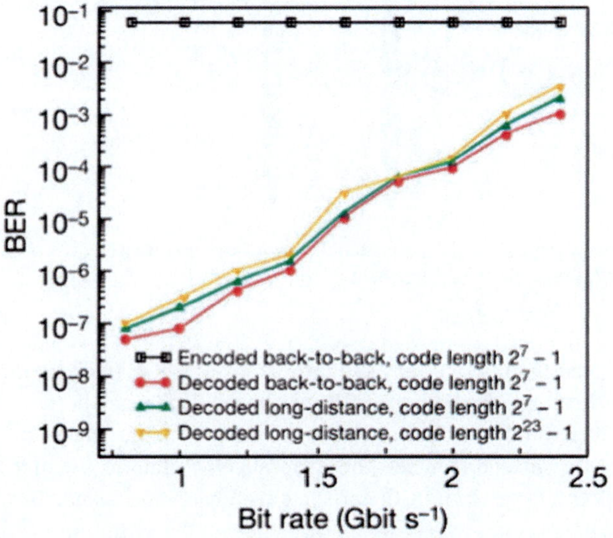

Fig. 4.13 Performances of chaos synchronization secure transmission using the metropolitan fiber network of Athens. Long-distance corresponds to a transmission over 120 km. BER of the encoded signal (squares), back-to-back decoded message (circles) and decoded message after transmission for two different code lengths (triangles). Courtesy of [32]

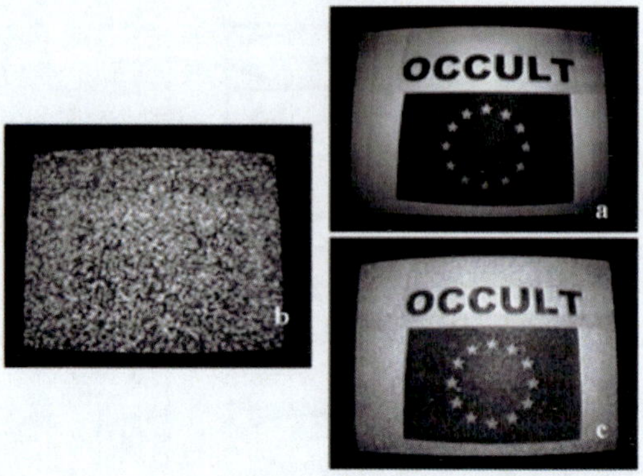

Fig. 4.14 Secure transmission of a composite video signal. **a** is a picture of the signal to be transmitted. **b** is what an eavesdropper retrieves by intercepting the encoded data stream and **c** is the decoded picture after subtraction. The deciphered picture is a little bit blurred compared to the initial one and a compromise has to be found between a high-quality transmission and a potential deciphering by an eavesdropper. Courtesy of [33]

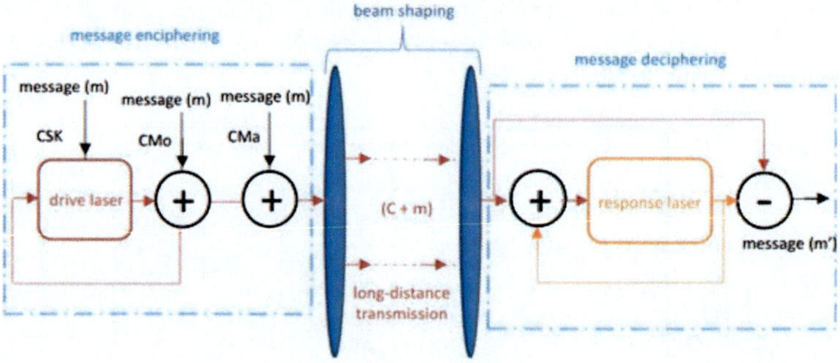

Fig. 4.15 Schematic diagram of the three encoding processes in the case of chaos communications in semiconductor lasers with optical feedback

or $x_2(t)$ is detected thanks to chaos synchronization. Thus, two chaotic options are offered for the drive and the response. However, the difference between two chaotic states with CSK method must be tiny, because otherwise, the message can be easily estimated from the attractors when the difference of chaotic fluctuations between the two states is too large. In non-linear systems, chaos synchronization is achieved through a finite delay time, which has an influence on the switching of chaotic states. Therefore, one must take into account the transient and finite response of signals for practical use of the systems.

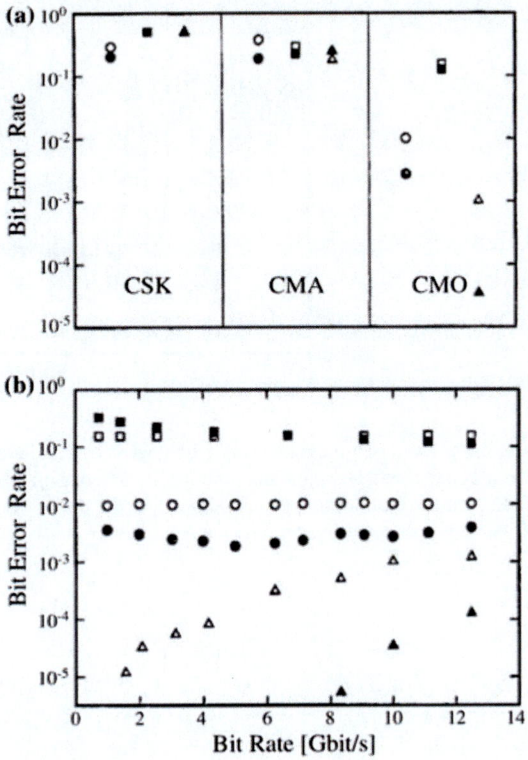

Fig. 4.16 Performance of BER for several chaotic configurations and several encoding processes. **a** BERs for optical injection system (marked as circles), optical feedback system (marked as squares), and optoelectronic feedback system (marked as triangles) of three different encryption schemes. The solid symbols mark the BER after the filtering, and the open symbols mark the BER before the filtering. **b** BER versus data transmission rate for the three previous configurations under CMo. Courtesy of [34]

4.1.3 Other Schemes to Communicate with Chaos

Other methods of communication relying on chaos have been successfully tested through an accurate control over the emitted chaotic fluctuations [37], as can be seen in Fig. 4.17. Control can be utilized to tune the signal emitted by the chaotic system so that it follows a prescribed symbol sequence, and this allows encoding any desired message from a chaotic oscillator. Most of the time, the phase diagram of such oscillators gathers multiple orbit attractors which are paired with a grammar [38]. The purpose is to choose a code whereby any message to be transmitted can be encoded with a sequence included in the grammar. This sequence must satisfy suitable constraints imposed by the dynamics in the presence of the control, which could be, for example, a periodic modulation. Therefore, it is of prime importance to control the spiking with thoroughness for reproducibility reasons and this can be achieved with drive-current modulation, external-cavity phase modulation or cavity-loss modulation [39], the perturbation being small compared to the chaotic fluctuations [40]. The previous chapter gave indications about control possibilities offered by the entrainment phenomenon. However, as the phase space breaks up into

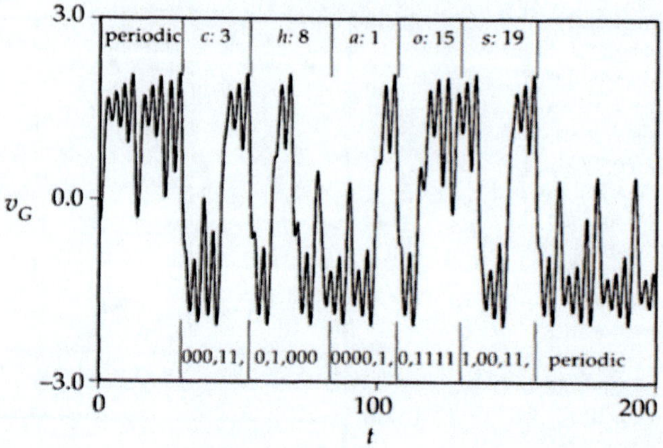

Fig. 4.17 Example of enciphering with chaos without involving synchronization. The word chaos is masked within the chaotic fluctuations. Each letter is shown at the top of the figure, along with its numerical position in the alphabet. Shown at the bottom are the corresponding binary code words. Extra bits (indicated by commas) are added to satisfy the constraints imposed by the grammar. Courtesy of [37]

distinct partitions [41], each of them corresponding to a distinct symbol, this solution is not necessarily the best in terms of privacy, even if efforts have already shown ways to improve it [42].

4.2 Secure Chaos Communications with Mid-Infrared QCLs

4.2.1 Free-Space Communications in the Mid-Infrared

Free-space communications have known a renewed interest in several optical domains where the atmosphere has a high transmission, as illustrated in Fig. 4.18 showing the solar and background radiations for a bunch of wavelengths. Experiments in the ultraviole have recently achieved a transmission at 1 Gbits/s with III-nitride micro-LEDs [43]. In the mid-infrared, several apparatus have emerged in order to produce high-bandwidth sources emitting in the transparency window around 4 μm. Communications up to 70 Mbits/s were demonstrated with interband cascade lasers (ICLs) [44] and transmissions at several Gbits/s were achieved with frequency down-conversion and up-conversion between 1550 nm and 3594 nm, respectively [45]. With QCLs, transmissions at similar high-speeds have been reported (Fig. 4.19b), both at room [46] and cryogenic temperatures with a free-space television link application [47]. In terms of secure communications, quantum cryptography systems with quantum key distribution (QKD) have been a candidate of choice with free-space sources emitting

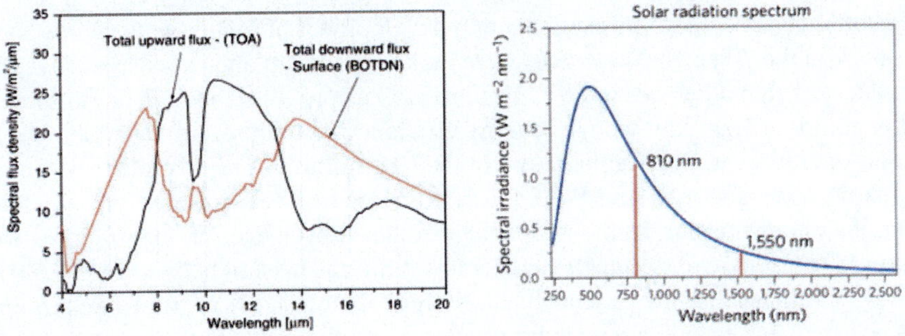

(a) (b)

Fig. 4.18 Background and solar radiations from long-wave infrared to ultra-violet light. **a** background from upward solar radiation and from downward thermal radiation. **b** Solar radiation spectrum from visible to near-infrared light. Courtesy of [52, 58]

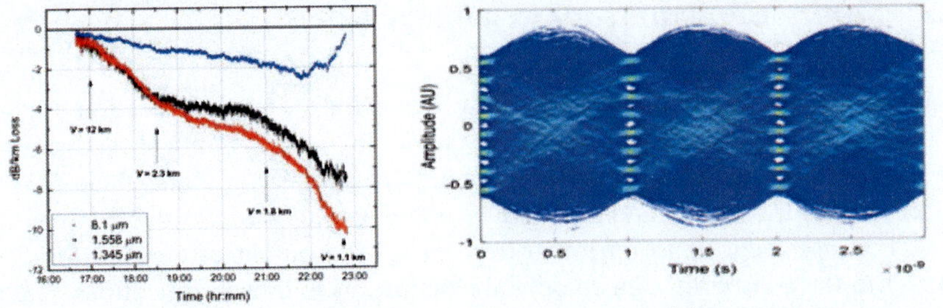

(a) (b)

Fig. 4.19 **a** Evolution of the transmission losses when varying the humidity of the atmosphere and thus, the visibility, for three wavelengths (8.1 μm, 1.558 μm and 1.345 μm). **b** Eye diagram of a state-of-the-art transmission with a mid-infrared QCL at room temperature. With a 8-level intensity modulation (PAM-8), the maximum data rate is 3 Gbits/s. Courtesy of [46, 57]

in the visible spectrum [48]. Yet, many efforts have tried to use these systems to communicate with satellites [49] but the background noise from the scattered sunlight, which is typically five orders of magnitude greater than the background noise during the night, complicates the process during daytime [50]. Therefore, many trials were done during the night [51] but this is not suitable for global applications where secure communications have to be required anytime. This is the reason why recent attempts have considered using near-infrared light to get rid of the perturbations without restraining to nighttime communications [52]. Overall, this technology is not versatile for every kind of secure communication applications and before QKD can be widely adopted, it faces a number of important challenges such as data transmission rate, distance, size, cost and ratio between the number of photons generated and the number of intricate photons finally available [53]. Nevertheless, it remains one of the best options in terms of privacy.

Despite the substantial investments into development of longer wavelengths systems back in the 70ies, these systems were bulky, inefficient and expensive [54]. The recent accelerated advances in QCLs [55] as well as progress in MCT photo-diodes [56] enable low size, low weight, energy efficient and therefore affordable optical systems operating at room temperature in the mid-infrared. The advantage of wavelengths between 3–5 μm and 8–11.5 μm compared to near-infrared wavelengths is that their transmission in the atmosphere remains high even though there is fog or haze [57], as shown in Fig. 4.19a. Moreover, a transmission between 8–11.5 μm implies stealth because the thermal black-body radiation shows a strong background at these wavelengths [58]. So it would be even more difficult to detect a secure transmission in this mid-infrared window and for the moment, only QCLs and CO_2 lasers can output powerful beams at such wavelengths.

4.2.2 Secure Communications with QCLs

The apparatus, described in Fig. 4.20, that we used for the synchronization of chaos is composed of a DFB master laser driven chaotic by external optical feedback and a DFB slave laser. Contrary to some schemes found in laser diodes (see [15] for instance), when the master and the slave laser are not coupled, the slave laser is not chaotic because there is no mirror in front of it. The output of the master is sent towards the slave laser through an optical isolator to avoid back-reflections, and the signal of the drive and the response are detected with a MCT detector. The signal from the MCT detectors are analyzed with a fast oscilloscope and a RF spectrum analyzer. Apart from the chaotic features observed with the RF spectrum analyzer and the digital oscilloscope, it is possible to confirm that the master signal is well injected into the slave by observing the behavior of the optical spectrum with a FTIR. Indeed, under free-running operation, the slave laser is mono-mode and it turns to be multi-mode when the synchronization is effective, as can be seen in Fig. 4.21.

For secure communications, the purpose is to have chaotic fluctuations with the broadest RF spectrum possible because this means that the maximum frequency of the message to be transmitted, as well as the privacy of the process, can be increased. Following the conclusions from the previous chapter, the first idea would be to try to synchronize chaos obtained when the QCL is pumped with a quasi-continuous wave because it gave the best results in terms of chaos complexity. However, this configuration may not be suitable because it can be difficult to include the message to be transmitted while the laser is already pumped with a QCW and furthermore, the transmission tentative can become easier to detect for an eavesdropper, thus destroying the benefit of the stealth. Also, slight changes in temperature occur when operating the laser in QCW mode. If, on the one hand, the master laser is driven chaotic in this configuration and, on the other hand, the slave laser is pumped with a continuous bias, it is not possible to synchronize the chaotic fluctuations of the master for the entire time period of the QCW, as shown in Fig. 4.22. This is due to the wavelength detuning because temperature shifts and this is a fundamental

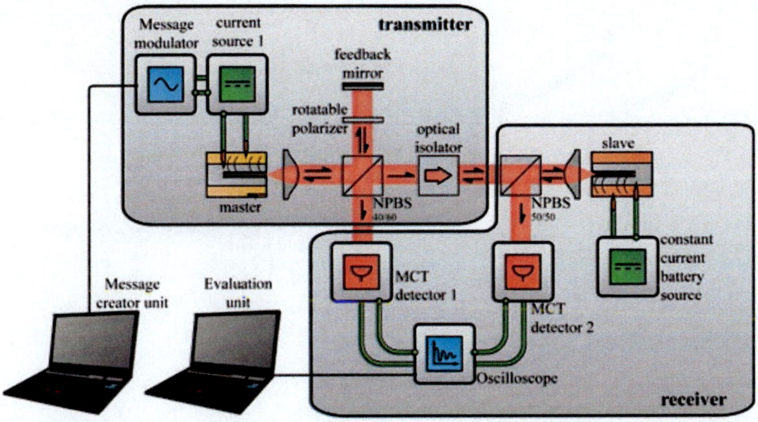

Fig. 4.20 Setup for chaotic secure communications with QCLs. The apparatus is split into two parts: one is dedicated to chaos masking and transmission and the other is dedicated to synchronization and recovery. NPBS: non-polarizing beam-splitter, MCT: Mercury-Cadmium-Telluride

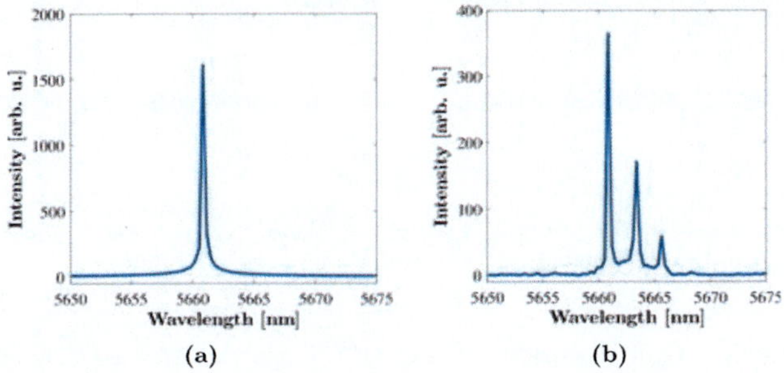

Fig. 4.21 **a** Optical spectrum of the free-running slave laser when biased at 512.4 mA. **b** Optical spectrum of the slave laser after injection of the output of the master laser when the slave laser is biased at 512.4 mA

result for QCLs which are proved to be highly temperature dependent. Indeed, we experimentally observed that the master laser and the slave laser can be synchronized only if they have the same wavelength, with an uncertainty of a few hundreds of MHz. The synchronization is also effective when the master QCL has the same wavelength than one of the side-modes of the slave QCL, and vice versa. However, it seems that the best case for synchronization occurs when the DFB peak of the master coincides with that of the slave. Consequently, it is not possible to take advantage of the chaos studied with a quasi-continuous bias because it detrimentally degrades the synchronization. It is interesting to note that we did not try to apply a QCW on both the slave and the master QCL to see if this could improve the overall quality of the chaos synchronization. As it is possible to have islands of chaos with a CW pumping, we investigated the possibility of chaos synchronization and secure communication

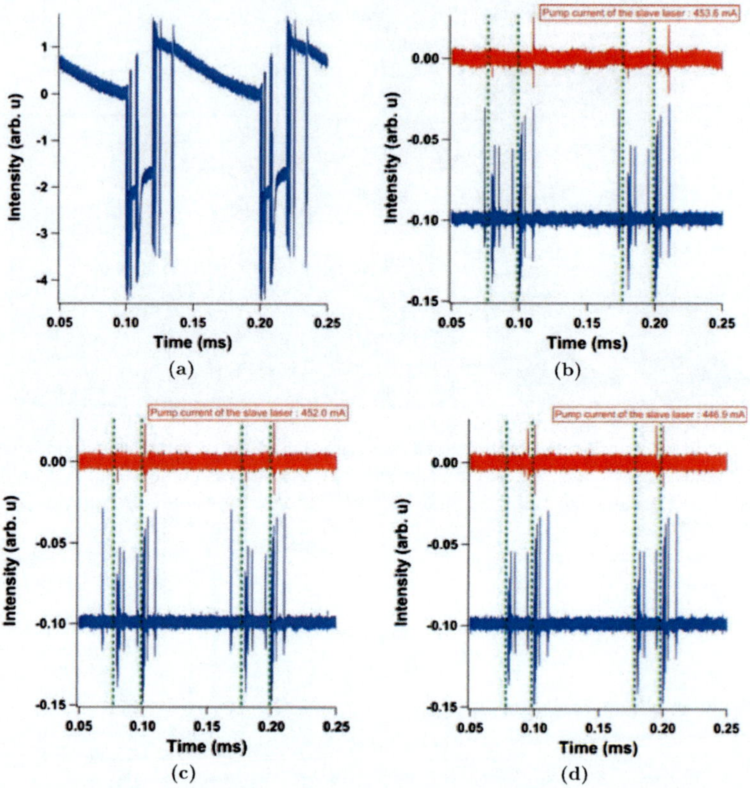

Fig. 4.22 Synchronization trial when the master QCL is pumped with a QCW bias and driven chaotic. **a** Output of the chaotic master laser pumped with a QCW before low-frequency filtering. **b–d** Slave QCL is pumped with several continuous biases. In each figures, the green dashed lines indicates the step limits of the 80 % duty cycle QCW, which are filtered by the detector. Depending on the pump current of the slave laser, only parts of the slave output is synchronized with the chaotic fluctuations of the master. The synchronization in the slave output is characterized by red spikes which are less numerous than blue spikes. This shows that a total synchronization cannot be achieved with a continuous bias for the slave laser if the master is pumped with a QCW bias

with this configuration. Also, this allows comparisons with the pioneering studies dealing with other semiconductor lasers.

So far, only numerical efforts have proven the possibility of a secure free-space communications but in the near-infrared, with laser diodes [59]. When the optical frequency of the master and the slave are matched, synchronization occurs. Modifying some of the parameters can also lead to anti-synchronization, as aforementioned, though it was not clear in our case what are the parameters leading to anti-synchronization in lieu of synchronization. As the best results in terms of mid-infrared secure transmission were achieved with anti-synchronization, we focus on this specific case (see in Fig. 4.23). As previously explained, several techniques can

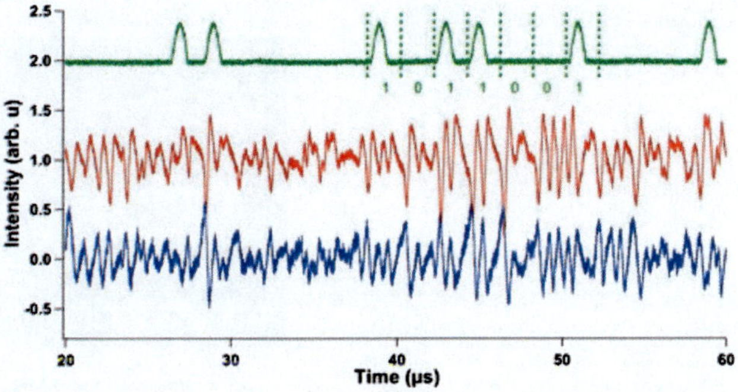

Fig. 4.23 Timetraces of the intensity of the chaotic master laser (red), the slave laser (blue) and the encoded message (green) which is magnified by a factor 30 to be clearly observed. The slave laser is anti-synchronized with the master laser and the message is well hidden within the chaotic carrier. The binary translation is given for a piece of the encoded message. A '1' is coded with a half-sine then a blank, while a '0' is coded with two consecutive blanks

be used to hide a message within a chaotic carrier. Because we performed the communications with two QCLs, the message has to be included directly in the bias current. The voltage amplitude of the enciphered bit code is thirty times lower than that of the largest chaotic fluctuations. The frequency coding and pattern was chosen accordingly with the cut-off frequency of the modulated current source (low-pass filter below 2 MHz) and that of the MCT detector (high-pass filter above 1 MHz). This is the reason why a square pattern could not be selected, because many frequencies included in the pattern would have been suppressed, either by the current source or by the detector. In order to comply best with these constraints, a half-sine pulse pattern at 1 MHz was chosen because it is composed of very few frequencies outside the frequency range 1–2 MHz. Nonetheless, this was not enough to ensure that the bit is not distorted or creates an undershoot, so a blank always follows a '1'-bit. To sum up, a '1' is encoded with a half-sine and a subsequent blank, while a '0' is encoded with two consecutive blanks, and an example of coding is shown in Fig. 4.23 (green curve).

Once the message is hidden and the anti-synchronization is achieved, the quality of the transfer is evaluated with correlation diagrams (Fig. 4.24). Then, the reconstruction of the message $M(t)$ is realized by computationally evaluating the difference intensity of the chaos transmitter $I_{ct} = |E_{master}(t) + M(t)|^2$ and the intensity of the chaos-synchronized slave $I_{slave} = |E_{slave}(t)|^2 \simeq |E_{master}(t)|^2$ resulting in $I_1 - I_2 = 2\Re(E_{master} \cdot M(t)) + |M(t)|^2$. Since the dynamics of the chaotic field E_{master} is much faster than the message $M(t)$, a low pass filter is applied to the difference and finally, $M(t)$ is obtained. The experimental result of the recovery can be found in Fig. 4.25, as well as the eye diagram for each component in Fig. 4.26. The eye diagram is a tool to evaluate the quality of the transmission, and a successful communication is assessed by an open eye. As already explained with the example of

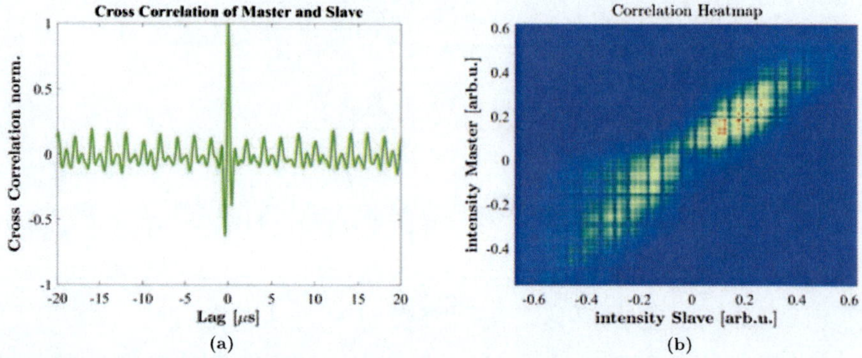

Fig. 4.24 1D (**a**) and 2D (**b**) correlation diagrams for the intensity of the slave and the master QCL. The signal of the slave laser was flipped so that the 1D correlation diagram has a maximum value of 1 and the 2D correlation diagram shows a positive correlation (instead of –1 and negative correlation, respectively)

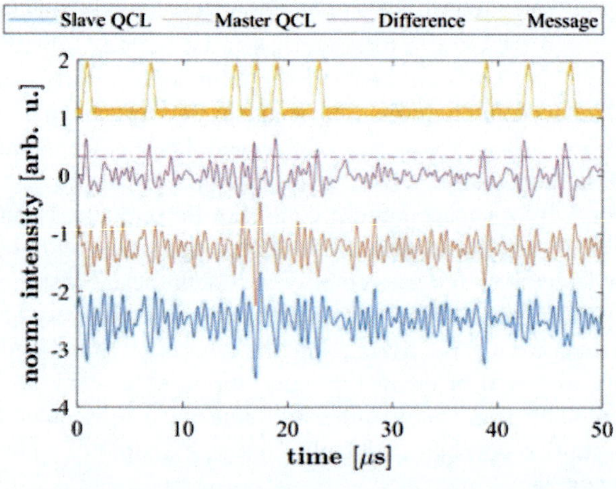

Fig. 4.25 Recovery process to decode the enciphered message. The blue curve is the filtered intensity of the slave QCL. It is also flipped so that it can be easily compared with the filtered intensity of the master laser (in red). The violet curve represents the subtraction between the red curve and the blue curve and the dashed line corresponds to the threshold used to detect a '1'-bit. The yellow curve is the magnified initial message, subsequently hidden within the red curve. Among the 9 '1'-bits present in the yellow timetrace, 8 are successfully recovered

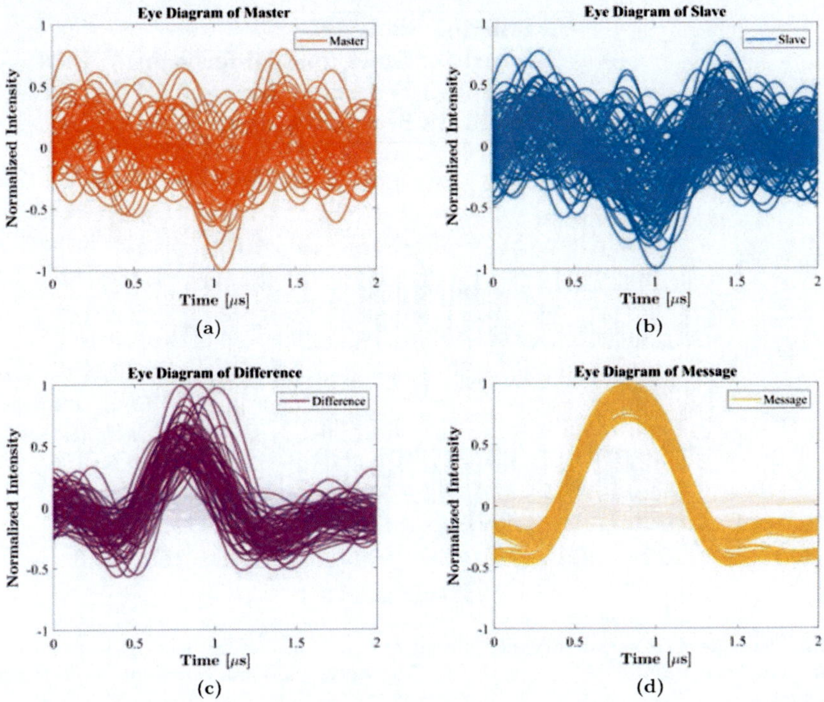

Fig. 4.26 Experimental eye diagrams for the 4 components of the previous figure. From these diagrams, one can see that it is impossible to recover the message only from the signal of the master QCL but it becomes possible for most of the bits from the difference signal. The eye diagram of the initial message shows some undershooting, hence justifying why a '1'-bit must be followed by a blank. Bits deciphered as '0' are drawn with a light color while bits deciphered as '1' are drawn with a stressed color (particularly visible in the difference eye diagram where the '1' and '0' are well separated)

the fiber secure transmission, a compromise has to be found between a transmission with a low BER and an impossible recovery for a potential eavesdropper. The best BER we were able to achieve, while ensuring a camouflaged message, is 6% (12 errors out of 191 bits in Fig. 4.27). It was possible to achieve lower values of BER but the message started to become obvious among the eye diagram of the master laser and consequently, the privacy of the transmission was threatened. Indeed, Fig. 4.28 shows that the red diagram displays a repeated pattern and, even if the message was not clearly visible in the chaotic timetrace, this singularity in the eye diagram could give indications to a potential eavesdropper.

Several opportunities can be envisioned in order to enhance the performances of this experimental proof-of-concept. The first one will be to increase the distance between the master and the slave laser because it is currently only one meter long. If a distance of several kilometers can be achieved, secure communications can be well adapted for transmissions between ground stations or vehicles and various aircrafts.

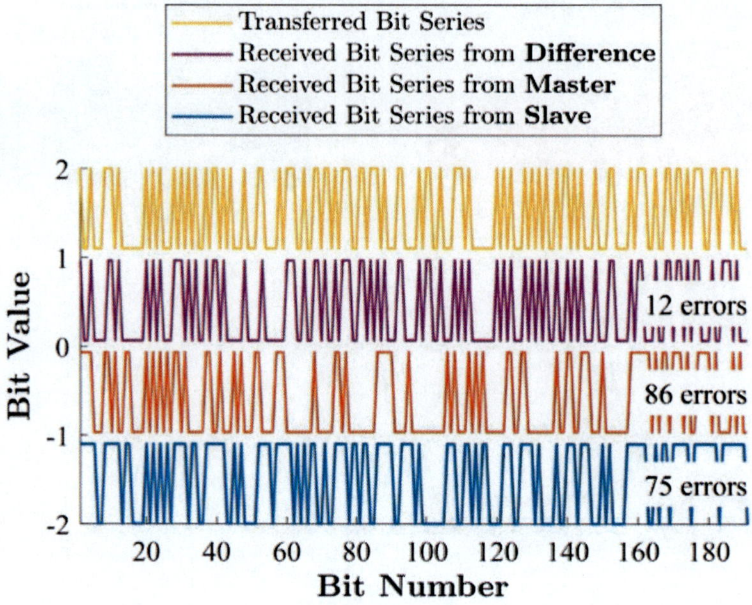

Fig. 4.27 Number of errors and associated bit values when deciphering the message by comparison with the initial message with 191 bits (in yellow). The secure mid-infrared transmission achieves a BER of 6%

Examples of transmissions are sketched in Fig. 4.29. To perform such long-distance communications, it is mandatory to switch from the current 5.7 μm wavelength towards wavelengths that are more suitable for optimal transmissions through the atmosphere, for instance 4 μm and 10 μm. While the latter can almost solely be achieved with QCLs, the former can be output in other semiconductor lasers like ICLs. The best option between these two lasers could be the one that is able to generate chaos with the largest bandwidth. The data rate of the hidden message is also relevant and even if we have recently achieved a 600 MHz modulation in a QCL by using a bias-tee, we may force the laser at a higher frequency in the case of ICLs. However, a transmission rate of dozens of Gbits/s can sometimes be required and this does not seem realistic with the aforementioned semiconductor lasers. The best option in this case would be to transmit only a secret key with the chaos secure communication and then use another means to achieve the very high-speed transmission. As the key can be as small as a few dozens of kbits [60], the data rate is not an issue because we showed it was possible to reach nearly 1 Mbits/s but in this case, chaotic secure communication is challenged by QKD. In the following, we will show promising methods based on optical injection to increase the chaotic bandwidth of a QCL.

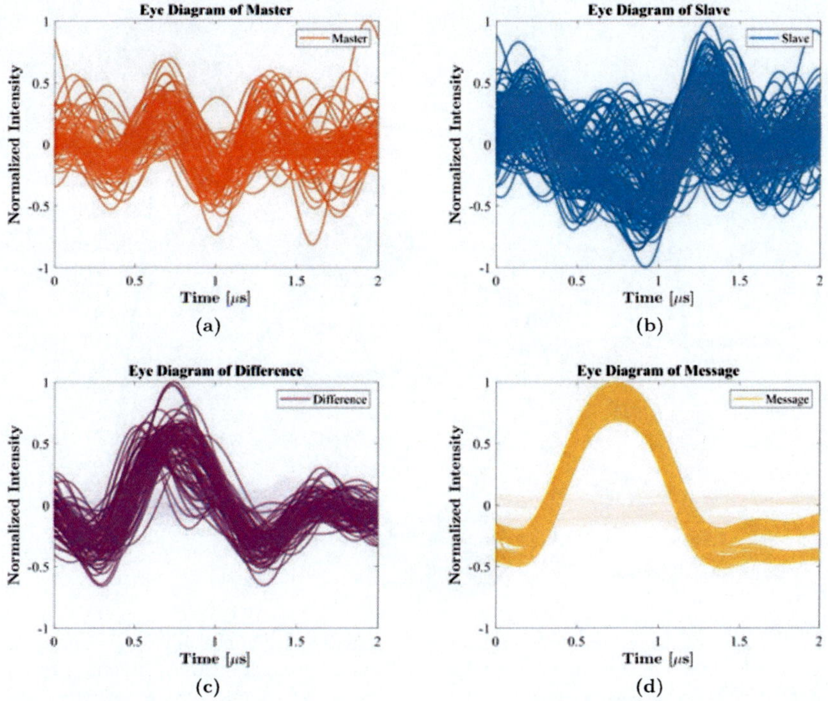

Fig. 4.28 Experimental eye diagrams for a case where the amplitude of the hidden message is too large and thus the message starts to be detectable within the chaotic carrier. A recurrent pattern appears in the eye diagram of the master laser, which was not the case in the previous configuration. In this case, the BER is 2% after recovery and subtraction

4.3 Improving Chaos Maximum Frequency in QCLs with Optical Injection

4.3.1 Benefits of Optical Injection

Similarly to what has been described for chaos synchronization, optical injection consists in injecting the light of a master laser into a slave laser. However in this case, the master laser is not driven chaotic. In the optical injection experimental setup described in Fig. 4.30, an optical isolator between the master and slave laser is a key element to avoid optical feedback of the master after reflection on the slave facet, or optical injection of the slave inside the master. Without isolator, the two lasers impact each other, it corresponds to mutually coupled lasers, which also experience complex dynamics [61]. We tested a mutual coupling scheme in the case of QCL chaos synchronization and saw that, even when the coupling between the two lasers is weak, the slave QCL can really disturb the non-linear pattern of the master QCL, hence justifying the necessity of an optical isolator between the two lasers. This is

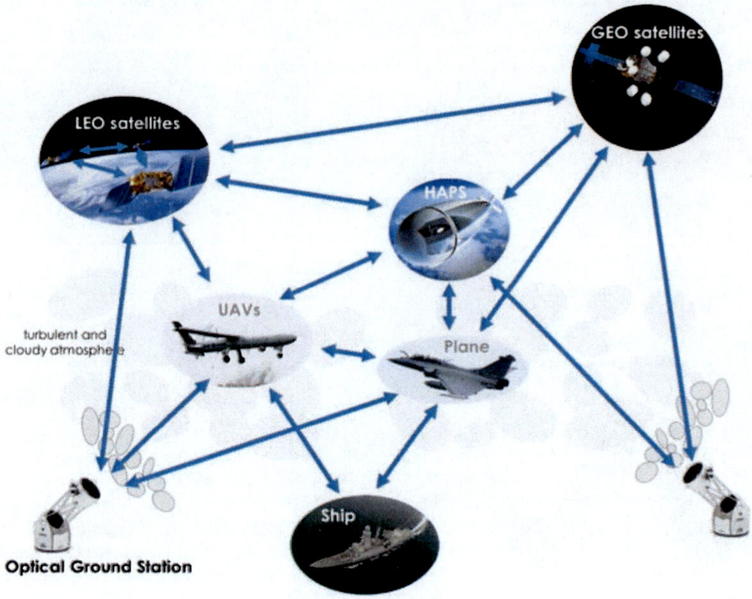

Fig. 4.29 Sketch of the several possibilities for secure free-space communications, from ground to space level. UAV: unmanned aerial vehicle, HAPS: high-altitude platform station, LEO: low earth-orbit, GEO: geostationary orbit

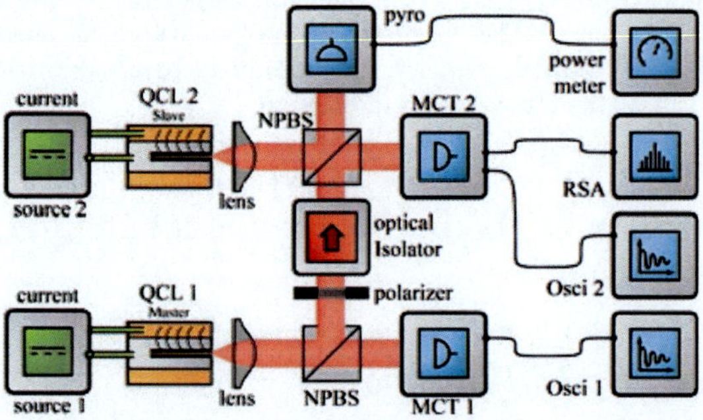

Fig. 4.30 Setup for the injection experiment. The apparatus is similar to that used for chaos synchronization, except that the master laser is not driven chaotic here. NPBS: non-polarizing beam-splitter, MCT: Mercury-Cadmium-Telluride, Osci: digital oscilloscope, RSA: electric spectrum analyzer

illustrated by Fig. 4.31. Optical injection can improve the emission properties of the slave laser with a significant impact. The case of injection-locking has already been mentioned and, under these conditions, the emission properties of the slave laser are improved. For instance, the side-modes of a FP-slave laser can be suppressed [62] while the spectral linewidth of the slave laser can be reduced by one order of magnitude [63]. Other improvements have been studied, both numerically and

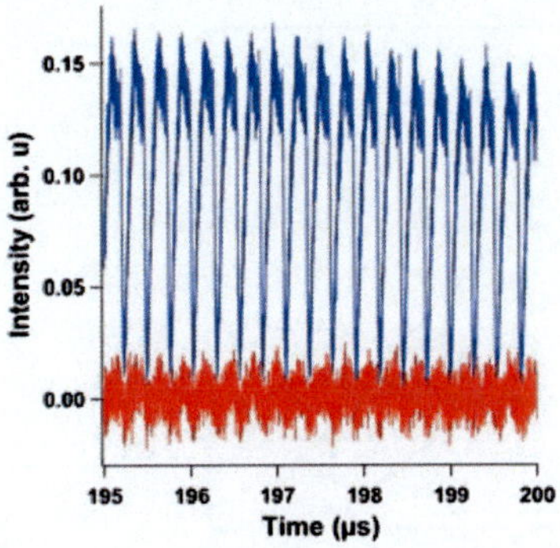

(a)

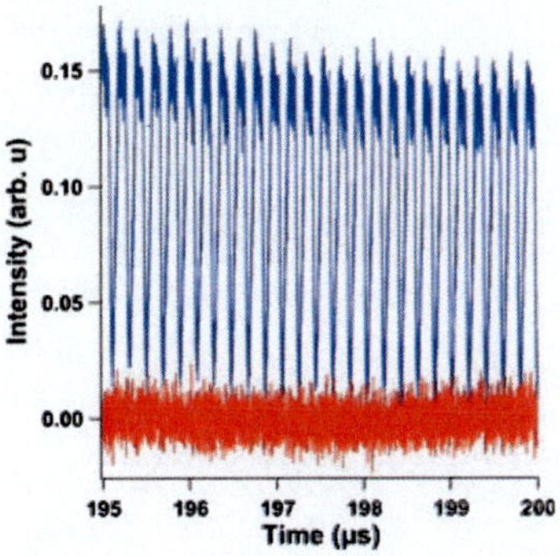

(b)

Fig. 4.31 **a** The master laser (in blue) is driven chaotic and is weakly coupled into the slave laser (in red) which weakly synchronizes with the chaotic fluctuations of the master. There is no optical isolator between them, so the slave laser influences in turns the master output. **b** When hiding the path between the two lasers, the synchronization is (obviously) lost but the frequencies of the master chaos are also modified. This stresses that an optical isolator is required between the two lasers, otherwise the slave can disturb the master laser, even if they are weakly coupled

experimentally, in terms of intensity, frequency noise and bandwidth. If the master laser has low intensity and phase noise, then the slave laser tends to reproduce the intensity and frequency noise of the master [64]. For the RF bandwidth, it was shown that a high injection rate (up to 45%) drastically enhances the bandwidth of the slave laser [65], as illustrated in Fig. 4.32. This paves the way for communications with faster data transmissions, without being limited by the frequency of the relaxation oscillations of the laser [66]. In laser diodes, optical injection was also found useful in enhancing the chaos bandwidth of a slave laser subject to external optical feedback [67]. If the master is strongly injected (via an optical isolator) towards the slave laser already driven chaotic because of EOF, the non-linear dynamics is not changed. For instance, the chaotic fluctuations do not turn into period-1 oscillations, but the bandwidth of the observed chaos is strongly increased. This technique, explained in Fig. 4.33, is also very relevant in terms of secure communications if a third laser is used for the synchronization of the enhanced chaos [68]. We did not intend to reproduce that scheme with QCLs even if it seems very promising for lasers in which the chaos maximum frequency is below the relaxation oscillation frequency.

Taubman and co-workers [69] were the first to report injection-locking of a DFB-QCL into a second DFB-QCL. By gradually increasing the master laser power, they observed an increase of the locking range, up to ± 500 MHz for the maximum achievable injection ratio of 3%. By studying the heterodyne beating between the slave and master lasers, a clear reduction of the slave linewidth was also observed. Finally, optical injection-locking was used to suppress residual amplitude modulation (RAM). Direct modulation of the bias current of a QCL indeed produces both frequency (FM) and amplitude modulations (AM), the latter being a source of noise in the case of FM spectroscopy. Under optical injection-locking, the RAM was reduced by 49 dB and

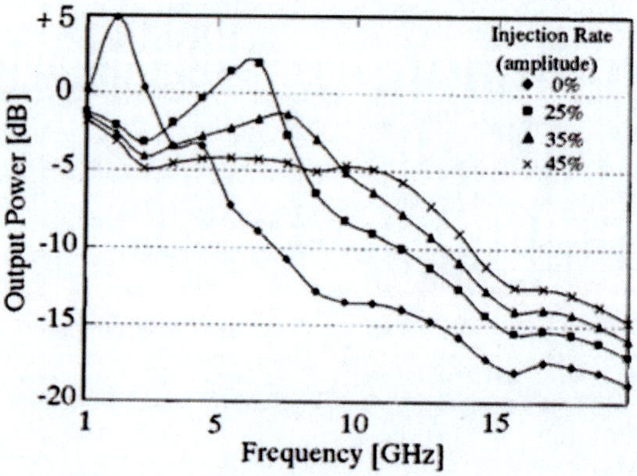

Fig. 4.32 Numerical result of bandwidth enhancement by a strong optical injection in laser diodes. The bias currents for the master and slave laser are the same, and the stronger the injection, the larger the bandwidth of the slave laser. Courtesy of [65]

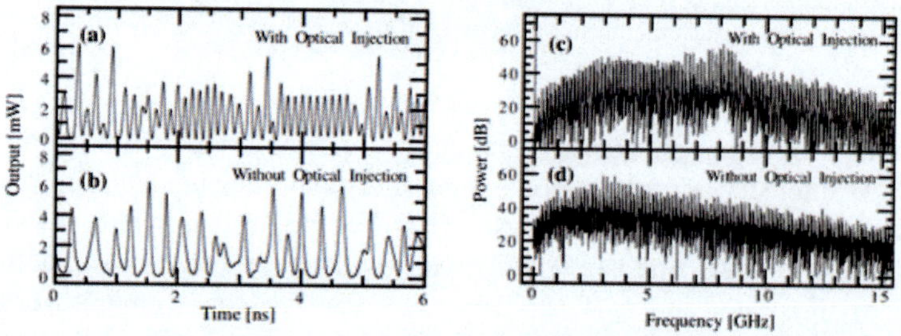

(a) (b)

Fig. 4.33 Numerical study of the influence of strong injection on the chaos bandwidth in a laser diode. **a** and **c** with optical injection, **b** and **d** without optical injection. The bandwidth is increased with optical injection, thus paving the way for using it as a chaos generator. Courtesy of [67]

could even be totally suppressed by applying an additional modulation on the slave bias current, out of phase with the one applied on the master laser. The RAM reduction using optical injection was also studied numerically [70]. Optical injection-locking can also be exploited to reduce the noise of a QCL. By injecting a DFB-QCL into another one, the relative intensity noise (RIN) at low frequency can be reduced by 20 dB/Hz [71]. Experimentally, under the same conditions, a maximum RIN decrease of around 10 dB/Hz was observed [72]. When tuning the master laser frequency over the whole slave laser range, the otherwise well-suppressed side-modes of the slave QCL will lock successively, inducing RIN reduction and output power increase. By injecting a frequency comb into a FP-QCL, the low-frequency frequency noise was also significantly reduced [73]. Up to 100 kHz, the slave QCL will reproduce the behavior of the master, and therefore decrease its frequency noise by 3–4 orders of magnitude. As a consequence, the linewidth of the laser is also reduced from a few MHz down to 20 kHz. A few theoretical studies, based on the QCL rate equations, also focused on the improved modulation properties of a QCL under optical injection [74–76]. All these studies concluded that no unstable locking region would exist in QCLs, due to the ultra-fast carrier lifetime. However, no experimental studies on the modulation of a QCL under optical injection-locking have been performed so far to verify these predictions, although the modulation bandwidth has been significantly improved up to 14 GHz in a QCL under radio-frequency (RF) injection-locking [77].

4.3.2 Rate Equations Under Optical Injection and Analytical Model

Similarly to the case of optical feedback, when neglecting spontaneous emission, if the slave QCL is under an electric field $E = \sqrt{S}e^{i\phi}e^{i\omega_{SL}t}$ and $E_{inj} = \sqrt{S_{inj}}e^{i\omega_{ML}t}$ is the electric field of the master laser, the rate equations of the QCL subject to optical injection can be written as:

$$\frac{dN_3}{dt} = \eta\frac{I}{q} - \frac{N_3}{\tau_{32}} - \frac{N_3}{\tau_{31}} - G_0\Delta NS \tag{4.7}$$

$$\frac{dN_2}{dt} = \frac{N_3}{\tau_{32}} - \frac{N_2}{\tau_{21}} + G_0\Delta NS \tag{4.8}$$

$$\frac{dN_1}{dt} = \frac{N_3}{\tau_{31}} + \frac{N_2}{\tau_{21}} - \frac{N_1}{\tau_{out}} \tag{4.9}$$

$$\frac{dS}{dt} = \left(N_{pd}G_0\Delta N - \frac{1}{\tau_p}\right)S + 2k\sqrt{S_{inj}S}\cos(\phi) \tag{4.10}$$

$$\frac{d\phi}{dt} = \frac{\alpha}{2}\left(N_{pd}G_0\Delta N - \frac{1}{\tau_p}\right) - \Delta\omega - k\sqrt{\frac{S_{inj}}{S}}\sin(\phi) \tag{4.11}$$

where N_j is the carrier density of level j, with $\Delta N = N_3 - N_2$, η is the conversion efficiency, I is the bias current. τ_{ij} corresponds to the carrier lifetime from level i to level j, τ_{out} is the characteristic time for the electron to tunnel into the injector, τ_p is the photon lifetime inside the laser cavity. G_0 corresponds to the net modal gain for one period, N_{pd} to the number of periods and α to the linewidth enhancement factor. $\Delta\omega$ is the frequency detuning, defined as $\Delta\omega = \omega_{ML} - \omega_{SL}$. Finally, k is the injection coefficient, defined as:

$$k = \frac{1}{\tau_{in}}2C_l \tag{4.12}$$

where τ_{in} is the internal-cavity round-trip time and C_l is the external coupling coefficient as defined in Eq. 3.15 or 3.16.

From Eqs. 4.7–4.11, the analytical model for a QCL under optical injection has been developed by Erneux and co-workers [78]. The rate equations are normalized with respect to the photon lifetime τ_p, with $s = t/\tau_p$ the normalized time, and defining the new set of variables (Y, Y_{inj}, Z, V), with Y and Y_{inj} the normalized amplitude of the slave and master laser electric fields, respectively, Z the normalized carrier density difference, and V the normalized carrier density of level $|2\rangle$, by:

$$\begin{aligned} Y &= G_0 \tau_{32} S \\ Y_{inj} &= G_0 \tau_{32} S_{inj} \\ \Delta N &= \frac{1+2Z}{N_{pd} G_0 \tau_p} \\ N_2 &= \frac{1}{N_{pd} G_0 \tau_p} \frac{1+V}{\frac{\tau_{32}}{\tau_{21}} - 1} \end{aligned}$$

Other parameters, such as the normalized lifetimes γ_1, γ_2, γ_3 and γ_4, the normalized bias current P, the normalized injected power γ and the normalized detuning Ω are expressed as:

$$\begin{aligned} \gamma_1 &= \frac{\tau_p}{\tau_{32}} \\ \gamma_2 &= \left(\frac{\tau_{32}}{\tau_{21}} - 1\right) \frac{\tau_p}{\tau_{32}} \\ \gamma_3 &= \left(\frac{\tau_{32}}{2\tau_{21}} - 1 - \frac{\tau_{32}}{2\tau_{31}}\right) \frac{1}{\frac{\tau_{32}}{\tau_{21}} - 1} \\ \gamma_4 &= 2\left(1 + \frac{\tau_{32}}{2\tau_{31}}\right) \\ P &= \frac{N_{pd} G_0 \tau_p \tau_{32}}{2} \frac{\eta (J - J_{th})}{q} \\ \gamma &= k \tau_p \sqrt{Y_{inj}} \\ \Omega &= \Delta\omega \tau_p \end{aligned}$$

4.3.3 Experimental Results with QCLs

This model for QCLs under optical injection has been thoroughly studied in a chapter of Ref. [79]. This subsection only focuses on the numerical conclusions resulting from this work and which can be compared with the experimental data retrieved when improving the maximum frequency of mid-infrared chaos. Depending on the injection rate and the detuning between the master QCL and the slave QCL, the latter exhibits damped oscillations, undamped oscillations or a pattern looking like the one found in LFF, but with a more periodic repetition rate. It also seems like the repetition

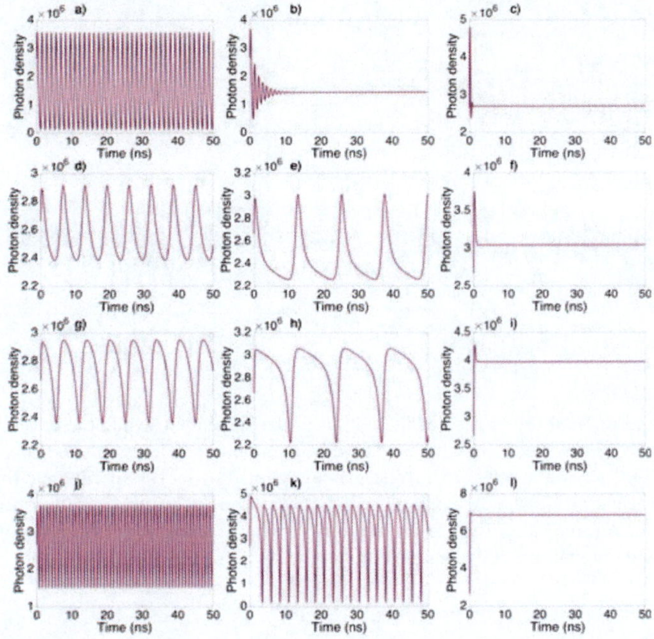

Fig. 4.34 Numerical time traces of the photon density for a FP-QCL subject to optical injection. The waveforms evolve with the frequency detuning and the feedback strength. First row: $\Delta f = +1.05$ GHz with **a** $k = -14$ dB, **b** $k = -13$ dB, **c** $k = -6$ dB. Second row: $\Delta f = +0.2$ GHz with **d** $k = -30$ dB, **e** $k = -27$ dB, **f** $k = -15$ dB. Third row: $\Delta f = -0.2$ GHz with **g** $k = -29$ dB, **h** $k = -26$ dB, **i** $k = -12$ dB. Fourth row: $\Delta f = -1.2$ GHz with **j** $k = -15$ dB, **k** $k = -8$ dB, **l** $k = -2$ dB. Courtesy of [79]

rate of the oscillations (if any) is dependent on the injection rate and its frequency is, at maximum, the value of the detuning. All these results are aggregated in Fig. 4.34.

Contrary to the numerical study, the experiments only focused on one injection strength. The non-linear phenomenon observed in Fig. 4.35 looks similar to the one retrieved with the numerical study for a negative detuning. That is to say, a LFF-like pattern with the main frequency centered around the value of the detuning. But even if this value is centered around the frequency detuning, the RF spectrum is quite large, with a bandwidth of roughly 50 MHz. So it is not easy to determine if what we observed was a periodic pattern as the one derived from the simulations or a deterministic form of chaos similar to what already described regarding LFFs. Further investigation is needed to derive the Lyapunov exponents of the retrieved signal and analyze if it is possible to synchronize this signal in a scheme with a third QCL. However, a recent study pointed out that it could be difficult to generate chaos solely with optical injection in QCLs [80]. In other semiconductor lasers, the behavior of the laser under injection varies a lot depending if it is considered as a highly-damped semiconductor laser [81] or not. In the case it is not, it is possible to observe several dynamical behaviors [82] from chaos to period-1 and period-2 oscillations, with

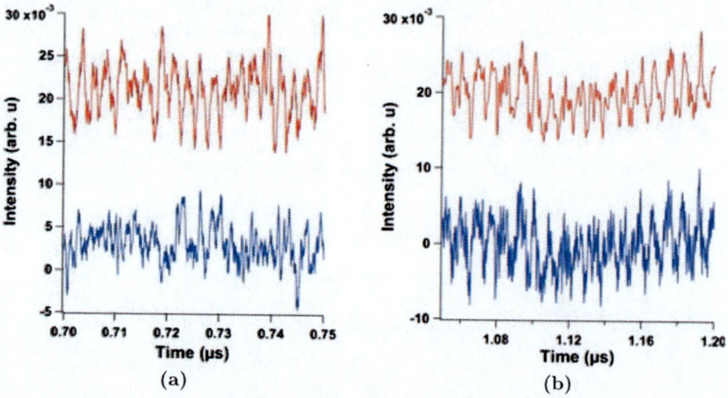

Fig. 4.35 **a** Experimental time traces of the output of the slave QCL when the bias current of the slave QCL is at 485.8 mA (red), showing a non-linear phenomenon close to the one observed in the simulation of Fig. 4.34 and at 485.5 mA (blue), showing only noise. **b** Experimental time traces of the output of the slave QCL when the pump current of the slave QCL is at 486.5 mA, after low-pass filtering (red) and before (blue). Even if the detuning is negative, the non-linear phenomenon does not seem reversed compared to the positive detuning case

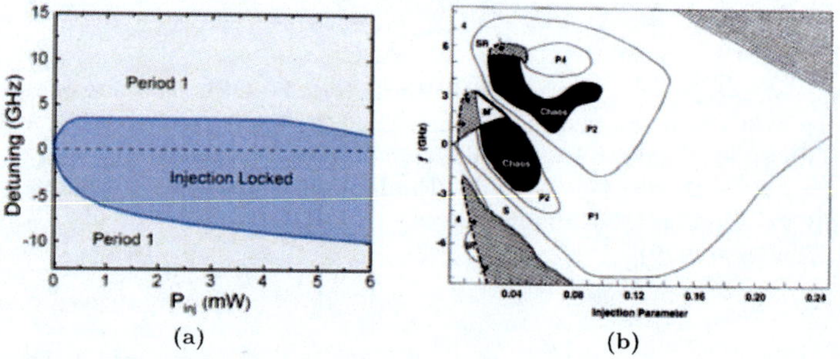

Fig. 4.36 **a** Injection map of a QD-laser, with only two areas depending on the injection strength and the frequency detuning. **b** Experimental mapping of the observed non-linear phenomena for a laser diode subject to injection. P1: period-1 oscillations, P2: period-2 oscillations, P4: period-4 oscillations, SR: stable regime. The behavior between the two types of lasers (highly-damped for the QD-laser contrary to the laser diode) is totally different. Courtesy of [81, 82]

sometimes stability islands, as shown in Fig. 4.36. It is worth noting that we did not experimentally observed an inversion of the pattern when switching from a negative detuning to a positive detuning. The pattern is visible from a near-zero detuning up to a detuning of 600 MHz, as can be seen in Fig. 4.37. The influence of the drive laser is strong for detuning values below 100 MHz and around 400 MHz. Further investigation is needed to understand why some frequencies are likely to have a stronger influence on the receiver laser. However it seems possible to increase the maximum frequency of chaos in QCLs up to a frequency of 400 MHz, provided that the observed phenomenon is deterministic chaos. The drawback is that the frequency of the pattern is very sensitive to the emission properties of the slave QCL or the

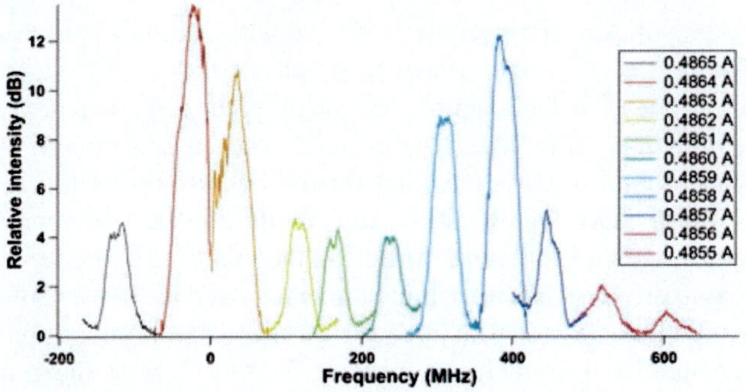

Fig. 4.37 RF spectrum of the slave QCL when varying the pump current of the slave QCL. The peak frequency for each case corresponds to the frequency detuning between the master laser and the slave laser. The influence of the master is visible up to a detuning of 600 MHz. Negative frequencies are only for convenience for the reader, and correspond to their positive frequency counterparts. The noise level is arbitrarily fixed at 0 dB

master QCL. For instance, if the bias current undergoes a 100 μA variation, this corresponds to a frequency detuning of 75 MHz and this modifies the frequency of the observed pattern. Even if not carefully assessed, a change in the heat sink temperature would lead to the same conclusion. A thorough experimental mapping is required to determine what are the most suitable regions in order to use injection as a chaos driver. Then, a third QCL would be required in order to synchronize the chaos previously generated and realize a secure communication, as described in the previous subsection.

4.4 Conclusion

By taking support on the previous experiments about optical chaos synchronization, we demonstrated a free-space synchronization and anti-synchronization at mid-infrared wavelength. This anti-synchronization was combined with message enciphering in order to realize a proof-of-concept of secure transmission, with a bit-rate of 0.5 Mbits/s and a BER of 6%. Even if the mid-infrared wavelength of this transmission makes it very relevant for further applications, improvements must be realized in order to comply with communications standards. Indeed, a transmission is considered error-free if the BER is lower than 4.7×10^{-3} with forward error correction (FEC) [83] and this is the value we are targeting. Even though CMo gave the best results in terms of BER with other semiconductor lasers, it would be interesting to compare this method with CMa and CSK, since BER is a parameter of prime interest in order to achieve a reliable communication.

Improvements of the bit-rate will also be possible by introducing the message through a bias-tee and not directly through the current source, as we did in the experimental proof-of-concept with the CMo scheme. Another option would be to optically modulate the QCL with a near-infrared laser [84]. Indeed, it is possible to quench the output of a QCL with a transient time of less than 0.1 ns in the case of strong injection [85]. Other schemes can also be envisioned, with a chaos generator not only composed by a feedback mirror, but also by an injector laser, as it was proven useful in chaos bandwidth enhancement in other semiconductor lasers. Another configuration also consists in injecting the chaos of a first laser (called the generator) into two different lasers, namely the master and the slave. The master will be electrically modulated in order to code the message that is subsequently masked by the chaos created by the synchronization of the master with the generator. The slave laser will also synchronize with the generator (not with the master) in order to generate chaos. Because the chaos created in the slave and the chaos created in the master come from the same injection process, it usually displays an improved correlation diagram and consequently leads to a better recovery of the message [86]. The combination of this method with the combined feedback/injection technique means that up to four similar QCLs may be necessary in order to reach the Graal of free-space chaotic secure communications which are suitable for every weather conditions thanks to their emitted wavelength in one of the transparency windows of the atmosphere. Overall, the privacy of the transmission will be assessed, not only with eye diagrams as we already showed, but also with state-of-the-art techniques relying on machine learning [87] and reservoir computing [88].

References

1. Pecora LM, Carroll TL (1990) Synchronization in chaotic systems. Physi Rev Lett 64(8):821
2. Chua LO (1992) The genesis of Chua's circuit. College of Engineering, University of California Berkeley, Electronics Research Laboratory
3. Chua LO (1994) Chua's circuit: an overview ten years later. J Circuits Syst Comput 4(02):117–159
4. Oliveira HM, Melo LV (2015) Huygens synchronization of two clocks. Sci Rep 5:11548
5. Ohtsubo J (2012) Semiconductor lasers: stability, instability and chaos. Springer, vol 111
6. Winful HG, Rahman L (1990) Synchronized chaos and spatiotemporal chaos in arrays of coupled lasers. Phys Rev Lett 65(13):1575
7. Roy R, Thornburg KS Jr (1994) Experimental synchronization of chaotic lasers. Phys Rev Lett 72(13):2009
8. Sugawara T, Tachikawa M, Tsukamoto T, Shimizu T (1994) Observation of synchronization in laser chaos. Phys Rev Lett 72(22):3502
9. Liu Y, De Oliveira P, Danailov M, Leite JR (1994) Chaotic and periodic passive Q-switching in coupled CO_2 lasers with a saturable absorber. Phys Rev A 50(4):3464
10. Haus HA, Mecozzi A (1993) Noise of mode-locked lasers. IEEE J Quantum Electron 29(3):983–996
11. Makarenko V, Llinás R (1998) Experimentally determined chaotic phase synchronization in a neuronal system. Proc Natl Acad Sci 95(26), 15 747–15 752
12. Shuai J-W, Durand DM (1999) Phase synchronization in two coupled chaotic neurons. Phys Lett A 264(4):289–297
13. Soriano MC, Garcia-Ojalvo J, Mirasso CR, Fischer I (2013) Complex photonics: dynamics and applications of delay-coupled semiconductors lasers. Rev Mod Phys 85(1):421

14. Ozaki M, Someya H, Mihara T, Uchida A, Yoshimori S, Panajotov K, Sciamanna M (2009) Leader-laggard relationship of chaos synchronization in mutually coupled vertical-cavity surface-emitting lasers with time delay. Phys Rev E 79(2):026210
15. Wedekind I, Parlitz U (2001) Experimental observation of synchronization and anti-synchronization of chaotic low-frequency-fluctuations in external cavity semiconductor lasers. Int J Bifurc Chaos 11(04):1141–1147
16. Boccaletti S, Pecora LM, Pelaez A (2001) Unifying framework for synchronization of coupled dynamical systems. Phys Rev E 63(6):066219
17. Rulkov NF, Sushchik MM, Tsimring LS, Abarbanel HD (1995) Generalized synchronization of chaos in directionally coupled chaotic systems. Phys Rev E 51(2):980
18. Voss HU (2000) Anticipating chaotic synchronization. Phys Rev E 61(5):5115
19. Boccaletti S, Kurths J, Osipov G, Valladares D, Zhou C (2002) The synchronization of chaotic systems. Phys Rep 366(1–2):1–101
20. Pecora LM, Carroll TL (1991) Driving systems with chaotic signals. Phys Rev A 44(4):2374
21. Cuomo KM, Oppenheim AV (1993) Circuit implementation of synchronized chaos with applications to communications. Phys Rev Lett 71(1):65
22. Colet P, Roy R (1994) Digital communication with synchronized chaotic lasers. Opt Lett 19(24):2056–2058
23. Sacher J, Elsässer W, Gobel EO (1991) Nonlinear dynamics of semiconductor laser emission under variable feedback conditions. IEEE J Quantum Electron 27(3):373–379
24. Mork J, Tromborg B, Mark J (1992) Chaos in semiconductor lasers with optical feedback: theory and experiment. IEEE J Quantum Electron 28(1):93–108
25. Annovazzi-Lodi V, Donati S, Scire A (1996) Synchronization of chaotic injected-laser systems and its application to optical cryptography. IEEE J Quantum Electron 32(6):953–959
26. Mirasso CR, Colet P, García-Fernández P (1996) Synchronization of chaotic semiconductor lasers: Application to encoded communications. IEEE Photonics Technol Lett 8(2):299–301
27. Vanwiggeren GD, Roy R (1998) Communication with chaotic lasers. Science 279(5354):1198–1200
28. Goedgebuer J-P, Larger L, Porte H (1998) Optical cryptosystem based on synchronization of hyperchaos generated by a delayed feedback tunable laser diode. Phys Rev Lett 80(10):2249
29. Sivaprakasam S, Shore K (1999) Signal masking for chaotic optical communication using external-cavity diode lasers. Opt Lett 24(17):1200–1202
30. Tang S, Liu J (2001) Message encoding-decoding at 2.5 Gbits/s through synchronization of chaotic pulsing semiconductor lasers. Opt Lett 26(23):1843–1845
31. Sanchez-Diaz A, Mirasso CR, Colet P, Garcia-Fernandez P (1999) Encoded Gbit/s digital communications with synchronized chaotic semiconductor lasers. IEEE J Quantum Electron 35(3):292–297
32. Argyris A, Syvridis D, Larger L, Annovazzi-Lodi V, Colet P, Fischer I, Garcia-Ojalvo J, Mirasso CR, Pesquera L, Shore KA (2005) Chaos-based communications at high bit rates using commercial fibre-optic links. Nature 438(7066):343
33. Annovazzi-Lodi V, Benedetti M, Merlo S, Norgia M, Provinzano B (2005) Optical chaos masking of video signals. IEEE Photonics Technol Lett 17(9):1995–1997
34. Liu J-M, Chen H-F, Tang S (2002) Synchronized chaotic optical communications at high bit rates. IEEE J Quantum Electron 38(9):1184–1196
35. Halle KS, Wu CW, Itoh M, Chua LO (1993) Spread spectrum communication through modulation of chaos. Int J Bifurc Chaos 3(02):469–477
36. Dedieu H, Kennedy MP, Hasler M (1993) Chaos shift keying: modulation and demodulation of a chaotic carrier using self-synchronizing Chua's circuits. IEEE Trans Circuits Syst II Analog Digit Signal Process 40(10):634–642
37. Hayes S, Grebogi C, Ott E (1993) Communicating with chaos. Phys Rev Lett 70(20):3031
38. Bollt E, Lai Y-C, Grebogi C (1997) Coding, channel capacity, and noise resistance in communicating with chaos. Phys Rev Lett 79(19):3787
39. Naumenko A, Loiko N, Turovets S, Spencer P, Shore K (1998) Chaos control in external cavity laser diodes using electronic impulsive delayed feedback. Int J Bifurc Chaos 8(09):1791–1799

40. Shinbrot T, Grebogi C, Yorke JA, Ott E (1993) Using small perturbations to control chaos. Nature 363(6428):411
41. Sukow DW, Gauthier DJ (2000) Entraining power-dropout events in an external-cavity semiconductor laser using weak modulation of the injection current. IEEE J Quantum Electron 36(2):175–183
42. Abel A, Schwarz W (2002) Chaos communications-principles, schemes, and system analysis. Proc IEEE 90(5):691–710
43. He X, Xie E, Islim MS, Purwita AA, McKendry JJ, Gu E, Haas H, Dawson MD (2019) 1 Gbps free-space deep-ultraviolet communications based on III-nitride micro-LEDs emitting at 262 nm. Photonics Res 7(7):B41–B47
44. Soibel A, Wright MW, Farr WH, Keo SA, Hill CJ, Yang RQ, Liu H (2009) Midinfrared interband cascade laser for free space optical communication. IEEE Photonics Technol Lett 22(2):121–123
45. Su Y, Wang W, Hu X, Hu H, Huang X, Wang Y, Si J, Xie X, Han B, Feng H et al (2018) 10 Gbps DPSK transmission over free-space link in the mid-infrared. Opt Express 26(26):34 515–34 528
46. Pang X, Ozolins O, Schatz R, Storck J, Udalcovs A, Navarro JR, Kakkar A, Maisons G, Carras M, Jacobsen G et al (2017) Gigabit free-space multi-level signal transmission with a mid-infrared quantum cascade laser operating at room temperature. Opt Lett 42(18):3646–3649
47. Martini R, Gmachl C, Falciglia J, Curti FG, Bethea CG, Capasso F, Whittaker EA, Paiella R, Tredicucci A, Hutchinson AL et al (2001) High-speed modulation and free-space optical audio/video transmission using quantum cascade lasers. Electron Lett 37(3):191–193
48. Buttler WT, Hughes RJ, Lamoreaux SK, Morgan GL, Nordholt JE, Peterson CG (2000) Daylight quantum key distribution over 1.6 km. Phys Rev Lett 84(24):5652
49. Nauerth S, Moll F, Rau M, Fuchs C, Horwath J, Frick S, Weinfurter H (2013) Air-to-ground quantum communication. Nat Photonics 7(5):382–386
50. Miao E-L, Han Z-F, Gong S-S, Zhang T, Diao D-S, Guo G-C (2005) Background noise of satellite-to-ground quantum key distribution. New J Phys 7(1):215
51. Hughes RJ, Nordholt JE, Derkacs D, Peterson CG (2002) Practical free-space quantum key distribution over 10 km in daylight and at night. New J Phys 4(1):43
52. Liao S-K, Yong H-L, Liu C, Shentu G-L, Li D-D, Lin J, Dai H, Zhao S-Q, Li B, Guan J-Y et al (2017) Long-distance free-space quantum key distribution in daylight towards inter-satellite communication. Nat Photonics 11(8):509
53. Diamanti E, Lo H-K, Qi B, Yuan Z (2016) Practical challenges in quantum key distribution. npj Quantum Inf 2:16025
54. McElroy JH, McAvoy N, Johnson E, Degnan J, Goodwin F, Henderson D, Nussmeier T, Stokes L, Peyton B, Flattau T (1977) CO_2 laser communication systems for near-earth space applications. Proc IEEE 65(2):221–251
55. Razeghi M (2020) High power, high wall-plug efficiency, high reliability, continuous-wave operation quantum cascade lasers at Center for Quantum Devices. In: Optical, opto-atomic, and entanglement-enhanced precision metrology II, vol 11296. International Society for Optics and Photonics, p 112961C
56. Kopytko M, Kębłowski A, Gawron W, Madejczyk P, Kowalewski A, Jóźwikowski K (2013) High-operating temperature MWIR nBn HgCdTe detector grown by MOCVD. Opto-Electron Rev 21(4):402–405
57. Corrigan P, Martini R, Whittaker EA, Bethea C (2009) Quantum cascade lasers and the Kruse model in free space optical communication. Opt Express 17(6):4355–4359
58. Puiu A, Fiorani L, Rosa O, Borelli R, Pistilli M, Palucci A (2014) Lidar/DIAL detection of acetone at 3.3 μm by a tunable OPO laser system. Laser Phys 24(8):085606
59. Annovazzi-Lodi V, Aromataris G, Benedetti M, Merlo S (2008) Secure chaotic transmission on a free-space optics data link. IEEE J Quantum Electron 44(11):1089–1095
60. Waks E, Inoue K, Santori C, Fattal D, Vuckovic J, Solomon GS, Yamamoto Y (2002) Quantum cryptography with a photon turnstile. Nature 420(6917):762
61. Erzgräber H, Lenstra D, Krauskopf B, Wille E, Peil M, Fischer I, Elsäßer W (2005) Mutually delay-coupled semiconductor lasers: mode bifurcation scenarios. Opt Commun 255(4–6):286–296

62. Kobayashi S, Kimura T (1981) Injection locking in AlGaAs semiconductor laser. IEEE J Quantum Electron 17(5):681–689
63. Mogensen F, Olesen H, Jacobsen G (1985) FM noise suppression and linewidth reduction in an injection-locked semiconductor laser. Electron Lett 21(16):696–697
64. Spano P, Piazzolla S, Tamburrini M (1986) Frequency and intensity noise in injection-locked semiconductor lasers: theory and experiments. IEEE J Quantum Electron 22(3):427–435
65. Ohtsubo J (2002) Chaos synchronization and chaotic signal masking in semiconductor lasers with optical feedback. IEEE J Quantum Electron 38(9):1141–1154
66. Meng XJ, Tong DT, Chau T, Wu MC (1998) Demonstration of an analog fiber-optic link employing a directly modulated semiconductor laser with external light injection. IEEE Photonics Technol Lett 10(11):1620–1622
67. Takiguchi Y, Ohyagi K, Ohtsubo J (2003) Bandwidth-enhanced chaos synchronization in strongly injection-locked semiconductor lasers with optical feedback. Opt Lett 28(5):319–321
68. Someya H, Oowada I, Okumura H, Kida T, Uchida A (2009) Synchronization of bandwidth-enhanced chaos in semiconductor lasers with optical feedback and injection. Opt Express 17(22):19 536–19 543
69. Taubman MS, Myers TL, Cannon BD, Williams RM (2004) Stabilization, injection and control of quantum cascade lasers, and their application to chemical sensing in the infrared. Spectrochim Acta Part A Mol Biomol Spectrosc 60(14):3457–3468
70. Yong KS, Haldar MK, Webb JF (2015) Theory of reduction of residual amplitude modulation in mid-infrared wavelength modulation spectroscopy by injection locking of quantum cascade lasers. IEEE J Sel Top Quantum Electron 21(6):115–124
71. Simos H, Bogris A, Syvridis D, Elsäßer W (2013) Intensity noise properties of mid-infrared injection locked quantum cascade lasers: I. Modeling. IEEE J Quantum Electron 50(2):98–105
72. Juretzka C, Simos H, Bogris A, Syvridis D, Elsäßer W, Carras M (2014) Intensity noise properties of mid-infrared injection locked quantum cascade lasers: II. Experiments. IEEE J Quantum Electron 51(1):1–8
73. Borri S, Galli I, Cappelli F, Bismuto A, Bartalini S, Cancio P, Giusfredi G, Mazzotti D, Faist J, De Natale P (2012) Direct link of a mid-infrared QCL to a frequency comb by optical injection. Opt Lett 37(6):1011–1013
74. Meng B, Wang QJ (2012) Theoretical investigation of injection-locked high modulation bandwidth quantum cascade lasers. Opt Express 20(2):1450–1464
75. Wang C, Grillot F, Kovanis V, Even J (2013) Rate equation analysis of injection-locked quantum cascade lasers. J Appl Phys 113(6):063104
76. Wang C, Grillot F, Kovanis VI, Bodyfelt JD, Even J (2013) Modulation properties of optically injection-locked quantum cascade lasers. Opt Lett 38(11):1975–1977
77. Renaudat St-Jean M, Amanti MI, Bernard A, Calvar A, Bismuto A, Gini E, Beck M, Faist J, Liu H, Sirtori C (2014) Injection locking of mid-infrared quantum cascade laser at 14 GHz, by direct microwave modulation. Laser Photonics Rev 8(3):443–449
78. Erneux T, Kovanis V, Gavrielides A (2013) Nonlinear dynamics of an injected quantum cascade laser. Phys Rev E 88(3):032907
79. Jumpertz L (2017) Nonlinear photonics in mid-infrared quantum cascade lasers. Springer
80. Zhao B-B, Kovanis V, Wang C (2019) Tunable frequency comb generation using quantum cascade lasers subject to optical injection. IEEE J Sel Top Quantum Electron 25(6):1–7
81. Hurtado A, Mee J, Nami M, Henning ID, Adams MJ, Lester LF (2013) Tunable microwave signal generator with an optically-injected 1310nm QD-DFB laser. Opt Express 21(9):10 772–10 778
82. Simpson T, Liu J, Huang K, Tai K (1997) Nonlinear dynamics induced by external optical injection in semiconductor lasers. Quantum Semiclassical Opt J Eur Opt Soc Part B 9(5):765
83. Chang F, Onohara K, Mizuochi T (2010) Forward error correction for 100 G transport networks. IEEE Commun Mag 48(3):S48–S55
84. Zervos C, Frogley M, Phillips C, Kundys DO, Wilson L, Hopkinson M, Skolnick M (2007) All-optical switching in quantum cascade lasers. Appl Phys Lett 90(5):053505

85. Chen G, Bethea CG, Martini R, Grant P, Dudek R, Liu H (2009) High-speed all-optical modulation of a standard quantum cascade laser by front facet illumination. Appl Phys Lett 95(10):101104
86. Yamamoto T, Oowada I, Yip H, Uchida A, Yoshimori S, Yoshimura K, Muramatsu J, Goto S-I, Davis P (2007) Common-chaotic-signal induced synchronization in semiconductor lasers. Opt Express 15(7):3974–3980
87. Pathak J, Lu Z, Hunt BR, Girvan M, Ott E (2017) Using machine learning to replicate chaotic attractors and calculate Lyapunov exponents from data. Chaos Interdiscip J Nonlinear Sci 27(12):121102
88. Pathak J, Hunt B, Girvan M, Lu Z, Ott E (2018) Model-free prediction of large spatiotemporally chaotic systems from data: A reservoir computing approach. Phys Rev Lett 120(2):024102

Chapter 5
Rogue Waves and Extreme Events

Rogue waves describe a phenomenon which was first observed in hydrodynamics, but whose denomination has been extended to many domains, from biology to optics as far as to finance. The particularity of rogue waves is to describe events with a tremendous characteristic (the height of the swell, the price of a market share, optical bursts...) occurring more often than they should. Another particularity of rogue waves is that they are unpredictable and for this reason, they are potentially hazardous. Consequently, they have been initially called rogue or freak waves because they were feared by sailors even if they were not officially reported and analyzed before the 20th century. Since then, rogue waves have raised interest because the analysis tools can be applied whatever the field of study. All that matters is to define, in each context, the quantity to compare to the threshold value.

After a brief review of the rogue wave phenomenon in contrasted areas, this chapter will focus on the optical rogue waves, and more specifically in semiconductor lasers. Several methods to observe rogue waves will be discussed with an emphasis on the external optical feedback method. This method gave rise to two different statistics of extreme events which will be thoroughly analyzed. Methods to trigger these events on-demand will also be explained, as optical rogue waves can be useful in sensing applications but are mostly found to be unpredictable.

5.1 Rogue Waves in Several Contexts

Rogue waves have long been considered a maritime legend among others like Krakens or Scylla and Charybdis. But following pictures taken by sailors in the 20th century, as the one shown in Fig. 5.1, they are now closely monitored because not only they exist but they can be harmful even for the largest tankers [1] and for deep-sea oil

O. Spitz, *Mid-infrared Quantum Cascade Lasers for Chaos Secure Communications*, Springer Theses, https://doi.org/10.1007/978-3-030-74307-9_5

Fig. 5.1 Pictures of the rarely observed rogue waves in the ocean. Courtesy of [1]

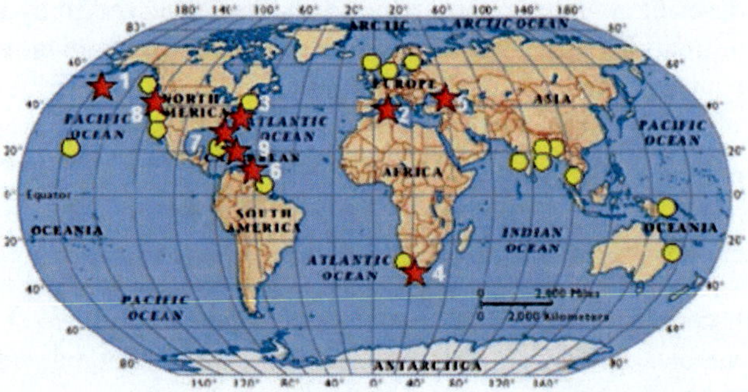

Fig. 5.2 Map of the confirmed oceanic rogue waves (stars) and other abnormally large waves observed in 2005. Courtesy of [3]

platforms [2]. Extensive maps, like the one in Fig. 5.2, are drawn every year and some areas, like the North Sea and the western part of the Atlantic Ocean, seem to be more prone to these gigantic waves [3]. The careful study of these maritime events started in the middle of the 20th century with the first records of the heights of the swell [4, 5]. Thorough efforts have then allowed to draw a global picture of the rogue wave phenomenon in the ocean [6], with recent papers still debating about their origin and occurrence rate [7, 8]. Before 2007 and the first observation of optical rogue waves, this term was exclusively utilized to describe water waves as the one monitored in Fig. 5.3. Following the discovery of events with similar properties but sometimes different mechanisms, the term 'rogue waves' has now been extended to a wide range of domains such as economics [9], acoustics [10], astrophysics [11] and biology [12], as illustrated in Fig. 5.4. The similarity with the oceanographic concept resides in the threshold criterion to take into account in order to consider an event as a rogue wave. Several criteria are thus generally applied to the data and the strictest one, leading to

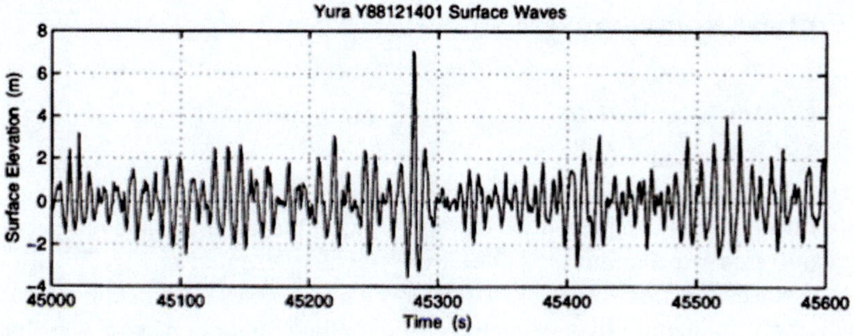

Fig. 5.3 Example of a rogue wave in the context of oceanography. This water wave was observed off Yura Harbor, Japan, in 1988. Courtesy of [6]

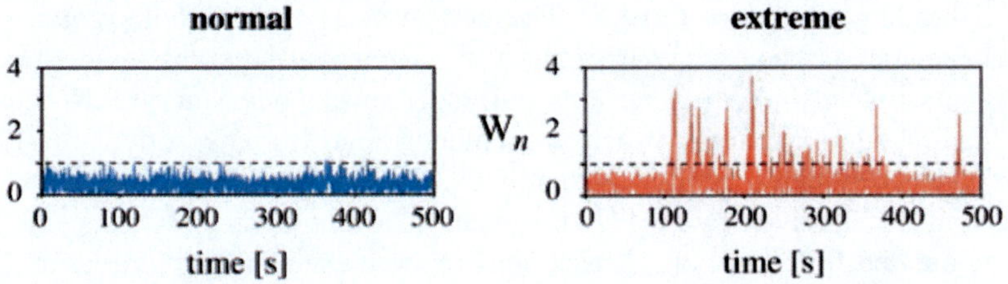

Fig. 5.4 Example of rogue waves in the context of biology. The diagram on the left displays no extreme event because it corresponds to a regular rat brain activity. The diagram on the right displays extreme events related to an epileptic crisis. The dashed line represents the threshold for extreme events. Courtesy of [12]

the identification of the lowest number of rogue waves, should be used. In order to detect rogue waves from experimental data, the probability density function (PDF) of an appropriate attribute of the data needs to be analyzed, and crest or wave heights are commonly used [13]. Once extracted from the data, heights are compared to a threshold value above which events can be considered as rogue. This term is applied for single waves that are extremely unlikely as judged by the Rayleigh distribution of wave heights [14]. In practice, this means that when one studies wave dataset of a finite length, a wave is considered to be a rogue wave if the wave height is higher than 2 times the significant height H_s of the dataset [3], or sometimes 2.2 times H_s [15], which represents the average of the higher third of heights. Another threshold widely used relies on the standard deviation (σ) and the mean value (μ) of the data and an event is considered as extreme if it exceeds $\mu + 8 \times \sigma$ [16]. However, a criterion with a lower threshold ($\mu + 6 \times \sigma$ or even $\mu + 4 \times \sigma$) is also commonly used without loss of generality [17, 18]. This threshold alone is however not sufficient to identify rare events and it is necessary to verify that the PDFs represent long-tailed distributions.

5.2 Optical Rogue Waves

5.2.1 Pioneering Works

Following the efforts about water rogue waves, the concept of optical rogue waves was first introduced in 2007 in the context of super-continuum generation in optical fibers at a near-infrared wavelength [19]. The experiments presented in Fig. 5.5 allowed observing pulses with extreme intensities and narrow timescales occurring rarely, yet much more frequently than expected based on the relatively narrow distribution of typical events. The statistical distribution of the events was indeed characterized by a L-shape showing a distortion of the regular gaussian distribution. Experimental findings were backed by numerical simulations relying on the generalized Non-Linear Schrödinger Equation (NLSE). The model was a powerful tool to confirm the soliton behaviour of the observed extreme events and it paved the way for the study of rogue events, not only in optical systems, but also in several physical systems because it was the first demonstration of such phenomenon apart from hydrodynamics.

The concept of rogue waves has then been extended to several optical domain, such as non-linear driven cavities [20], Raman fiber amplifiers and lasers [21, 22], pump-modulated fiber lasers [23], Kerr lens mode-locked Ti:Sapphire lasers [18], fiber ring lasers [24] and optically injected semiconductor lasers [16]. For the latter,

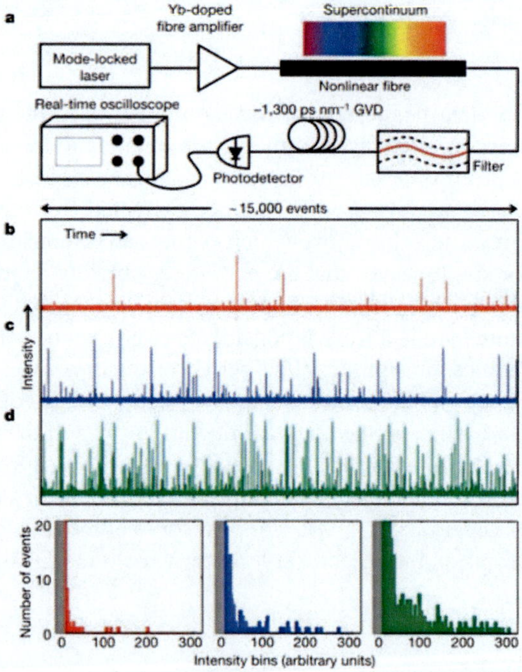

Fig. 5.5 **a** Experimental setup for observation of the first optical rogue waves. **b–d**, Time traces containing roughly 15,000 pulses each and associated histograms showing the specific L-shape for average power levels 0.8 mW (red), 3.2 mW (blue) and 12.8 mW (green), respectively. Only a few extreme events called rogue events reach intensities of at least 30–40 times the average value. These distributions are very different from those encountered in most stochastic processes. Courtesy of [19]

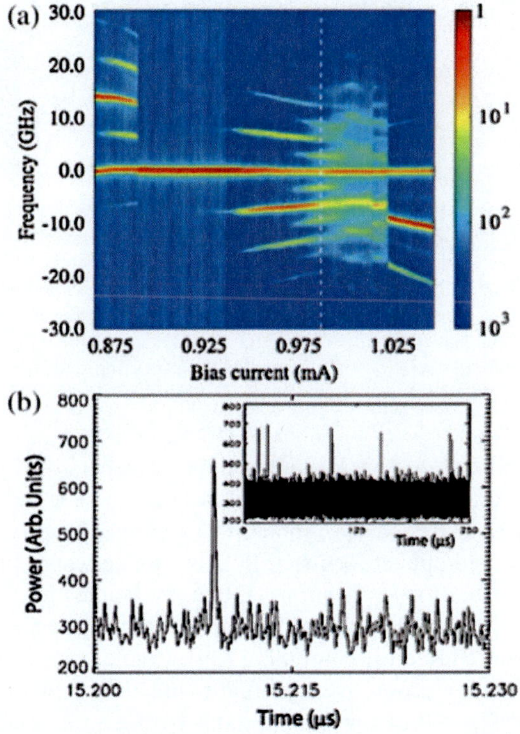

Fig. 5.6 First experimental proof of the rogue wave concept in a semiconductor laser under injection. **a** Experimental bifurcation diagram displaying the optical spectra of the slave laser versus the bias current. The dashed line denotes the zero detuning condition between the master and the slave lasers. **b** Typical time series of the slave laser intensity, where rare large pulses can be observed. Courtesy of [16]

the laser under study was a VCSEL biased below the limit of single transverse mode emission and its output wavelength was around 980 nm. Another similar laser was used for the injection process. Close to the zero detuning condition, the slave laser can generate giant intensity pulses, as illustrated in Fig. 5.6. A careful analysis of the probability density function confirmed a long-tailed distribution of these extreme pulses (Fig. 5.7), which can therefore be accounted as rogue waves. Simulations based on a simple noise-free rate equation model gave results in good qualitative agreement with the experiments, allowing for the interpretation of the sporadic high amplitude pulses as the result of a deterministic non-linear process.

5.2.2 *Extreme Pulses in Semiconductor Lasers with Feedback or Injection*

Most of the experimental and numerical works studying rogue waves and extreme events in semiconductor lasers focus on an injection scheme [16, 17, 25–31]. VCSELs and laser diodes are the most common devices in these experiments because they are renowned for their sensitivity to perturbations giving birth to non-linear phe-

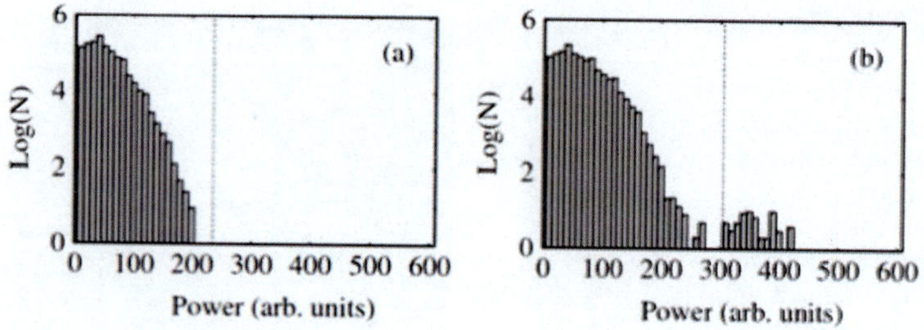

Fig. 5.7 Experimental histograms of the first optical rogue waves in a semiconductor laser, without rogue waves (**a**) and with rogue waves (**b**). N is the number of pulses detected in the time traces of the slave laser intensity. The vertical dashed lines denote the mean value plus eight times the standard deviation and thus set the limit for a pulse to be considered a rogue wave. Courtesy of [16]

nomena. A very limited number of papers deals with the study of extreme events in the case of optical feedback [32–34] and, among these papers, the conventional optical feedback is sometimes replaced by phase-conjugate feedback [35, 36] or filtered optical feedback [37]. As rogue waves pop up randomly, they are not very reliable for applications. Extensive studies have been carried out in order to take advantage of these sudden bursts for high precision remote-sensing [38] or optoelectronic reservoir computing [39]. Rogue waves can indeed be predicted in the case of the injection scheme because precursors are displayed in the time trace, as shown in Fig. 5.8. A few nanoseconds before a rogue event occurs, oscillations at a lower amplitude can be seen [26, 31], both in the photon intensity and in the optical phase. Similar numerical efforts were carried out in the case where rogue waves are generated with external optical feedback, and once again, a specific patterns occurring a few nanoseconds before the rogue events allows predicting them [32], as visualized in Fig. 5.9. These studies support a qualitative comparison of the behavior with feedback and with injection and, despite the differences in both systems, suggest that extreme intensity pulses in the optical feedback case could occur through a similar mechanism as in the optical injection case. It is also relevant to note that in both cases, the higher the threshold, the more pregnant the precursor before the advent of the rogue pulse.

As rogue waves can also be hazardous due to the sudden extra power they carry, several methods to suppress them have emerged. In the case of a laser diode subject to injection, oscillations are displayed during a few nanoseconds before the burst of the rogue wave during less than one nanosecond. In the case of feedback, another pattern is very likely to occur before an extreme event and the electrical signal can thus be anti-synchronized in advance [40] even if the timescales of the order of the nanosecond are quite challenging in the case of laser diodes. Similarly, an external modulation with an adequate range of frequency and amplitude parameters [17] or a high level of noise [26] can suppress extreme events in configurations which were supposed to be prone to extreme events.

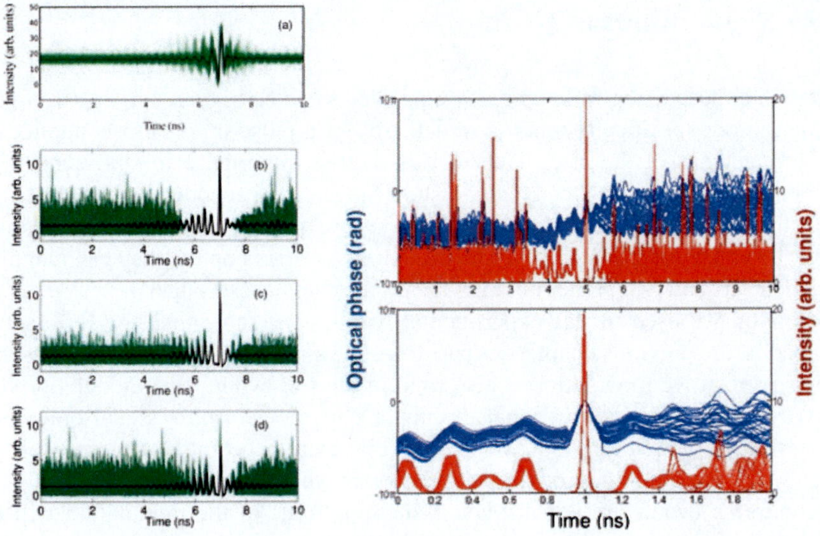

(a) (b)

Fig. 5.8 Illustration of the fluctuations preceding an optical rogue wave in the case of injection. The left column shows the superposition of 500 experimental time traces centered on rogue events (**a**) as well as a deterministic (**b**) and a stochastic simulation (**c**) with the $\mu + 8 \times \sigma$ threshold; **d** is the same configuration as (**b**) but with the $\mu + 4 \times \sigma$ threshold. The right column also shows a deterministic simulation and proves that the precursor can also be observed in the optical phase. Courtesy of [26, 31]

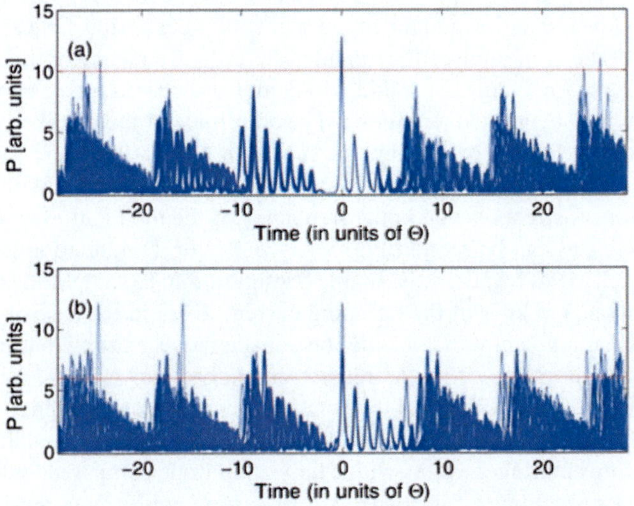

Fig. 5.9 Superposition time traces from a deterministic numerical model studying extreme pulses in the case of EOF. Time traces are centered at the peak of the pulse. The threshold for defining extreme pulses, indicated by the horizontal solid line, is equal to $\mu + 5 \times \sigma$ in (**a**) and is equal to $\mu + 3 \times \sigma$ in (**b**). In both panels the number of pulses is 52. Courtesy of [32]

5.2.3 Rogue Waves in QCLs

As we will expose in the following, extreme events can be triggered in a QCL under EOF, and more specifically when a modulation or a pulse-up pattern is applied to the electrical bias of the laser. Hence, it is of first importance to study and map this phenomenon because it can really disturb the message encoded via the electrical current of the QCL when it comes to realize a secure transmission. The succession of bits can be considered as a pulse-up and if extreme events pop up, they can magnify the hidden bits and reveal them in the viewpoint of an eavesdropper.

The setup we used in this experimental work is the same than the EOF setup described in the previous chapters. Apart from the sine-wave modulation, an asymmetric square-wave modulation is also used in the following. This modulation has a duty cycle of 20%, meaning that the upper part of the square signal only lasts for one fifth of the total period. We will see that extreme events are more likely to appear when the amplitude of the modulation is weak. Except for the next figure where extreme events are not displayed, the amplitude of the modulation will be limited to 5 mA, corresponding to less than 2% of the threshold current of the 5.6 μm QCL under study. In the following, we will focus on a rogue wave threshold corresponding to 2 times the significant height H_s of the dataset, which is equivalent to an abnormality index of 2.

As emphasized by the previous chapters, QCLs were proven to exhibit several non-linear dynamics usually found in other semiconductor lasers with external optical feedback. Further tuning of the spiking dynamics is possible through a sine-wave modulation of the pumping current. In this case, the chaotic dropouts synchronize with the periodic modulation and the number of spikes per period can vary through a modification of the amplitude of the modulation or the frequency of the modulation as we already saw. Analyzing the spiking dynamics showed that some configurations led to spikes with a magnified amplitude. The amplitude of these spikes is very sensitive to any change of the parameters, as shown in Fig. 5.10b where a modulation frequency of 2 MHz leads to spikes with an amplitude comparable with that of the sine modulation. A slight change in the frequency of the modulation annihilates the magnified spikes, as can be seen in Fig. 5.10a and c for a modulation frequency of 1.8 MHz and 2.2 MHz, respectively. In this configuration, the amplitude of the modulation corresponds to 28% of the pumping current. Even in the case of Fig. 5.10b, the spikes with a large amplitude cannot be considered as extreme because they do not meet the requirements of the aforementioned criterion. As extreme events are compared with the mean value and the standard deviation of the signal, decreasing the amplitude of the modulation or even suppressing the external modulation is one of the key features to trigger events with a large amplitude compared to the common values of the signal. Figure 5.11a shows a time trace with rogue waves when no external modulation is applied. This case differs from the usual configuration where EOF gives rise to deterministic chaos because the external cavity is asymmetric, or in other words, the feedback mirror is a bit tilted. The back-reflected light consequently hits the emission facet of the QCL in an uneven manner. Though this method is

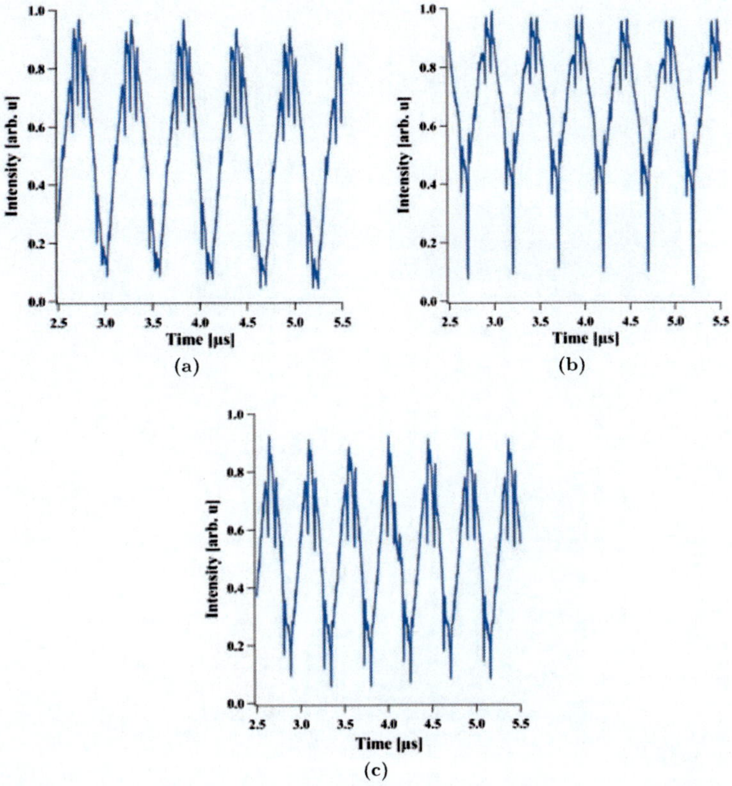

Fig. 5.10 Experimental time traces of the QCL output when external optical feedback is applied to the QCL and with a sine modulation of the continuous bias of: 1.8 MHz (**a**), 2 MHz (**b**), 2.2 MHz (**c**); All the waveforms were retrieved for a continuous bias of 430 mA and a modulation amplitude of 120 mA, but only the one in the middle exhibits dropouts with a magnified amplitude, which is however not enough to categorize them as extreme events

difficult to assess precisely, it seems to be mandatory in order to trigger events with a large amplitude without any external modulation. It is relevant to note that variations of the non-linear dynamics in diode lasers due to a mirror tilt were already studied more than thirty years ago. The authors had also envisioned new types of dynamics and potential applications [41].

Figure 5.11b shows that the time-interval between the events we observed is not easily foreseen, which is one of the characteristics of rogue waves. Other studies have shown that the separation time (S) between rogue events follows a Poissonian distribution [42] when the time elapsed between these events (numbered $k-1$ and k) is written in log scale: $S = log(T_k) - log(T_{k-1})$. A deviation of the statistics is sometimes found for high feedback ratios [35], and in this case, the curve is characterized by two slopes instead of one, as can be seen in Fig. 5.12a. In our case, Fig. 5.12b shows the evolution of the separation count in log scale with S

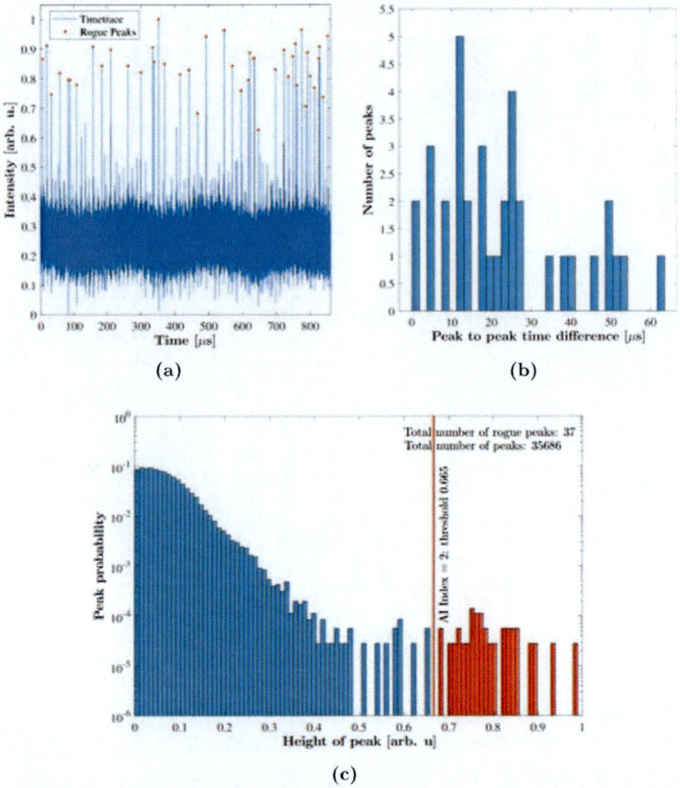

Fig. 5.11 **a** Experimental time trace with events meeting the rogue wave criterion. **b** Histogram of the time-intervals between extreme events. **c** Histogram of the retrieved maxima and the related threshold corresponding to 2 times the significant height H_s. Red bars gather the extreme events of the displayed time trace

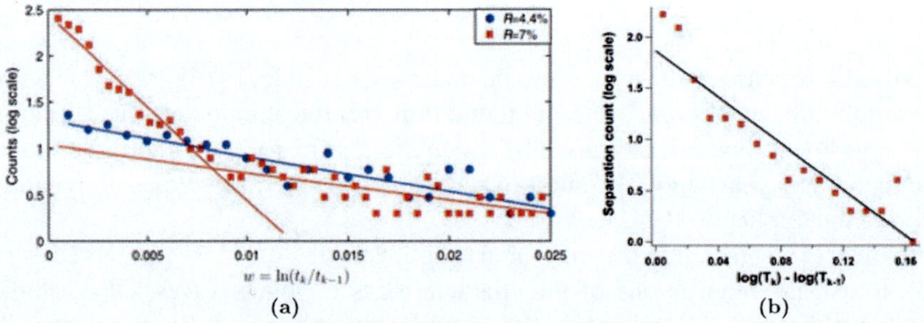

Fig. 5.12 **a** Experimental study of extreme events in a diode laser with phase-conjugate feedback. The ratio between the phase-conjugate feedback and the incident beam power can be tuned and lead to rogue waves for the higher values (above 3%). Statistical distribution of the time between extreme events for R = 4.4% (blue) and R = 7% (red). Courtesy of [35]. **b** Statistical distribution of the time-intervals between rogue events in log scale and the associated linear regression when aggregating 419 rogue waves intervals from several different time traces in the QCL configuration

when taking into account 419 rogue waves intervals obtained from many different time traces but with the same experimental conditions. The evolution is linear as encountered in other semiconductor lasers. Even if we are in a configuration with a very high feedback strength, a linear regression with one slope is in good agreement with our data, contrary to the result given in Ref. [35] for a high feedback strength.

The configuration without modulation of the pump current is thus not optimized for the triggering of extreme events. Indeed, it is not possible to predict when they will pop up and this prevents any possible utilization of these sudden bursts. Figure 5.11c gathers the statistic of the spikes found in the time trace for the aforementioned criterion. The duration of the record was chosen as a trade-off between the number of displayed rogue waves and a sufficient resolution to ensure that the retrieved events were not artifacts. There is a clear long-tail distribution in the statistics which is incompatible with a regular distribution of maxima and this means that events with a large amplitude appear more frequently than expected. This configuration is very similar to what happens when studying deterministic chaos in QCLs. Indeed, EOF can lead to low-frequency fluctuations but they are mostly unpredictable, which is not the case when a periodic external modulation allows synchronizing the spiking dynamics. Consequently, applying a periodic modulation with a low amplitude are one of the levers to display huge pulses while maintaining the entrainment phenomenon.

5.2.4 Dragon-King Events

The notion of dragon-king events was first introduced in 2009 [43] even if corresponding events have long been spotted [44]. Dragon-king have been observed in many different contexts, ranging from avalanches [45] to chaotic electronic oscillators [46] as far as to bursting neurons [47]. They are also characterized by giant events, but in this case, these events follow a power law except the occasional very large events which have a higher probability of occurrence, as illustrated in Fig. 5.13. One main characteristic of rogue waves distributions is that they are scale free, which means that events of arbitrarily large sizes are caused by the same dynamical mechanisms governing the occurrence of small and giant events. In contrast, giant pulses related to dragon-king events cannot be explained by a scale free distribution. This means that dragon-king events possess a distinct formation mechanism which may ease the prediction of these giant pulses [48].

As it is a sub-category of rogue waves, no study of the dragon-king events in optical systems have been reported so far. This does not mean that some of the aforementioned efforts on optical rogue waves do not partially contain dragon-kings among conventional rogue waves. In our study of rogue waves in QCLs, we have observed a quite rare configuration where giant pulses do not follow the conventional PDF (like the one in Fig. 5.11). In this configuration, giant pulses all have more or less the same amplitude and consequently, there is a gap in the PDF. As this case rarely occurred, we were not able to compute enough data to draw a PDF showing a local minimum and then a local maximum for a very large value of the amplitude.

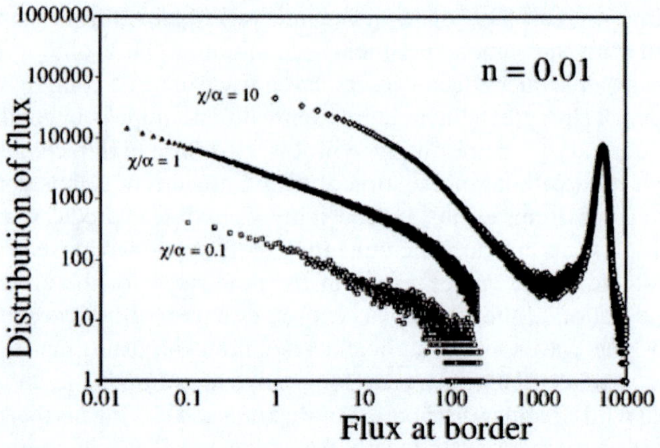

Fig. 5.13 Example of dragon-king events in the field of fluid mechanics. Distribution of the avalanches size for three different $\frac{\chi}{\alpha}$ parameters. For $\frac{\chi}{\alpha} = 10$, the probability density function exhibit a peak for very high values of flux at border. This distortion of the PDF is typical from dragon-king events (even though this name was not yet given in 1996). Courtesy of [44]

Figure 5.14 shows the timetrace with the potential dragon-king events and the related peak-to-peak histogram and PDF. The waveform looks very similar to the one in Fig. 5.11, except that no pulses with intermediate size occur.

5.2.5 *Predicting Rogue Waves in QCLs*

Some applications, such as improved remote-sensing or optical neuron-like systems, require extreme events to be predicted in order to take advantage of the extra optical power. Within other scientific areas, many efforts have also tried to forecast abnormal and striking events, for instance in medicine [49], finance [50], road traffic [51] or climate science [52]. Rogue waves and extreme events are also to be suppressed in cases where the surge can lead to the destruction of a component or when it can disrupt the transmission of a secure message. This inhibition process is also of prime interest for neuromorphic schemes in order to interrupt regular trains of bursts [53]. Methods to prevent extreme events to occur have been numerically tested for both the configuration with injection [26] and the configuration with EOF [32]. For instance, extreme events can be anticipated by studying precursors in the time traces. Figure 5.15a shows the temporal superposition for the 37 extreme events retrieved in Fig. 5.11. Two main indications can be seen in this figure when focusing on the average of the time traces: a dip occurs approximately 300 ns before the rogue event and this event is followed by an oscillation lasting a few microseconds with a decreasing amplitude. To draw a comprehensive comparison with the aforementioned numerical studies [26, 32], we also displayed the same superposition diagram but for

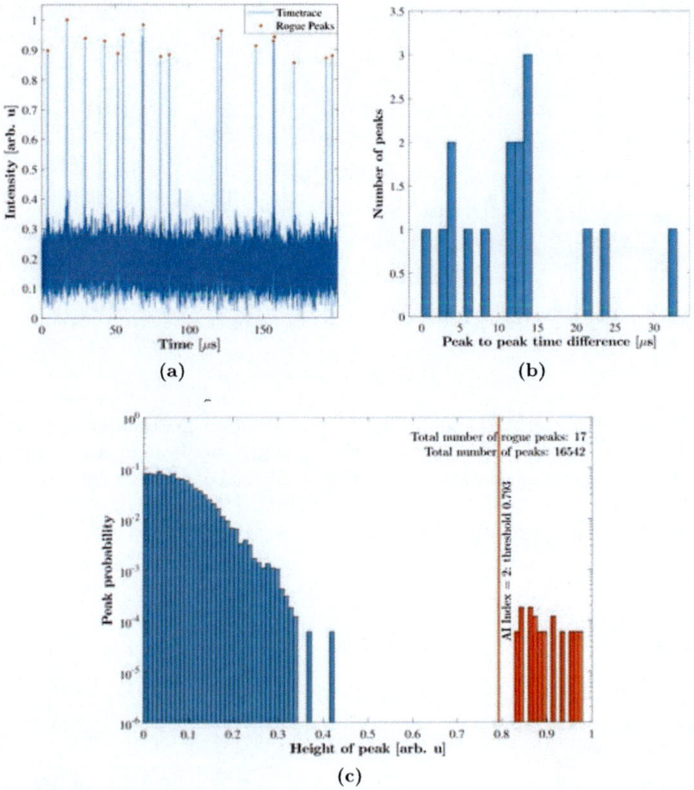

Fig. 5.14 **a** Experimental time trace with dragon-king events meeting the rogue wave criterion. **b** Histogram of the time-intervals between extreme events. **c** Histogram of the retrieved maxima and the related threshold corresponding to 2 times the significant height H_S. Red bars gather the extreme events of the displayed time trace

a higher threshold (abnormality $A = 2.4$) in Fig. 5.15b, taking into account 13 rogue peaks instead of 37. Contrary to the simulations of Figs. 5.8 and 5.9, increasing the threshold for giant pulses does not allow predicting more accurately rogue waves. However, the dip following the rogue events is more clearly drawn in the second case. The oscillation pattern following the rogue wave was globally more visible in the case where the threshold was lower. In the case of dragon-king events (Fig. 5.16), the dip following the giant event is still present but it is not really possible to extract an oscillation pattern as we did for the rogue waves. It is also difficult to be positive about the dip usually found before extreme events. In the case of dragon-king events, it seems to be there although secondary extreme events close to the main event disturb the superposition diagram. The overall relevance of the latter may be affected by the small number of events we were able to take into account and a dataset with more dragon-king events could allow us drawing a more precise conclusion. In the case of rogue waves, the superposition diagram from Fig. 5.15a shows that all the extreme

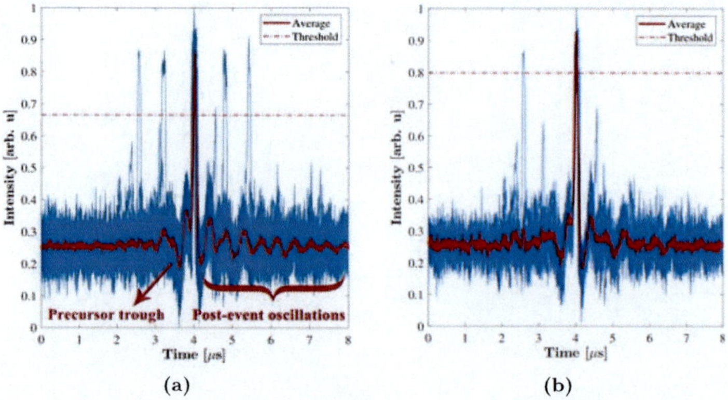

Fig. 5.15 **a** Time series centered on one single extreme event and superposition of 37 rogue waves. The corresponding averaged output power is displayed with a purple solid line and the threshold corresponding to an abnormality index of 2 is displayed with a dash-dot line; typical patterns occurring before and after the rogue wave are also underlined. **b** Same data plot but for 13 rogue waves surging above an abnormality index of 2.4

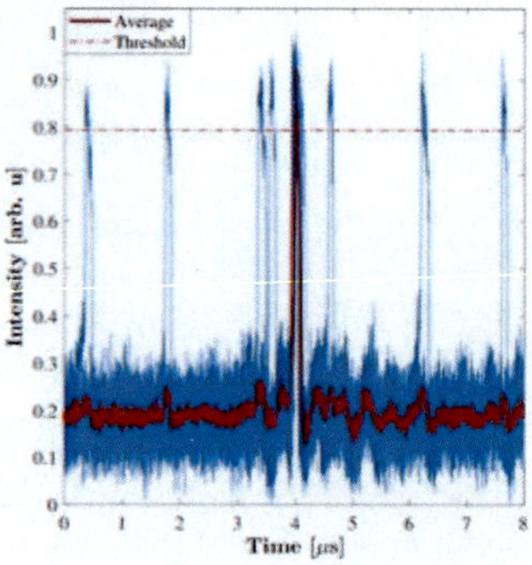

Fig. 5.16 Time series centered on one single extreme event and superposition of 17 dragon-king events. The corresponding averaged output power is displayed with a purple solid line and the threshold corresponding to an abnormality index of 2 is displayed with a dash-dot line

events (even those which are not that tremendous) can be predicted and potentially suppressed before they occur. We did not try to implement the latter method but the long elapsed time between the precursor and the rogue peak (a few hundred times larger than that observed with VCSELs or laser diodes) would ease such suppression in the case of unwanted rogue waves, as already envisioned in the case of a solid-state laser with electro-optical modulator [54]. We have to keep in mind that extreme events can affect the performances of QCLs and that methods have been successfully tested in other semiconductor lasers to get rid of them. An accurate cartography of the rogue wave phenomenon in QCLs would also be useful to determine the impact of several parameters such as the modulation frequency or the value of the spontaneous emission factor.

5.2.6 Controlling the Likelihood of Extreme Pulses

In this section, we focus on extreme events in QCLs. Extreme events may share statistical properties with rogue waves even if their behavior is different. They indeed both correspond to large deviations of the system state away from its nominal value but extreme events are characterized by a variation of a parameter. This variation leads to a deviation of the intensity statistics to the otherwise Gaussian distribution, hence explaining the occurrence of intense pulses.

Being able to generate extreme events in QCLs can be of prime importance for sensing applications. The mid-infrared windows is known to hold the fingerprint of many chemical molecules and QCLs has been successfully used for spectroscopic applications. However, rogue waves are not easy to predict less than one microsecond in advance, as we previously saw, and are consequently difficult to monitor for industrial applications. In this subsection, we explain how to experimentally trigger extreme pulses with a probability higher than 40%. Following the conclusions in the first subsection, applying a modulation with a low amplitude will decrease the mean value and standard deviation of the laser's output and may be prone to extreme events. Figure 5.17 shows the extreme events appearing when a sine modulation of 0.1 MHz at a low amplitude of 4 mA is applied to the QCL. It is important to note that, contrary to the CW rogue wave case, this configuration does not require a tilt of the feedback mirror. The waveform looks very similar to the one in Fig. 5.11a with extreme events seemingly distributed in a random manner. However, a more accurate analysis of the distribution shows that extreme events can only appear when the external sine modulation reaches a maximum. This is illustrated in Fig. 5.17b for two extreme events and this configuration shares similarities with the entrainment phenomenon case at 2 MHz (Fig. 5.10b) where the magnified events were only popping up for a specific phase-shift of the modulation signal. This result echoes the microsecond excitability process described in Ref. [55] for a laser diode with stochastic resonance, except that in our case, the amplitude of the extreme events is not consistent throughout the time trace. Even if this method allows forecasting the extreme events in a narrow time-interval around the extrema, the maximum success rate is very low (around 10%) and the global picture is that the triggering of these events is almost random. Furthermore, the amplitude of the extreme spikes is not constant throughout the time trace and this may not be suitable for applications like remote sensing. Sine modulation is however not the only lever to trigger extreme events in semiconductor lasers.

A recent numerical study showed that the best success rate for triggering such events was obtained when the excitation was a pulse-up signal [31] in the case of an injected laser diode. In this configuration, Jin and co-workers were able to achieve a success rate of 50%. We thus decided to implement a similar pulse-up configuration in the case of EOF by means of an asymmetric square-wave modulation. This square signal has a duty cycle of 20% and a low amplitude of 4 mA. The period of the pattern is 10 μs. Figure 5.18a shows a time trace of the QCL operated under the aforementioned conditions. Extreme events can only appear simultaneously with the

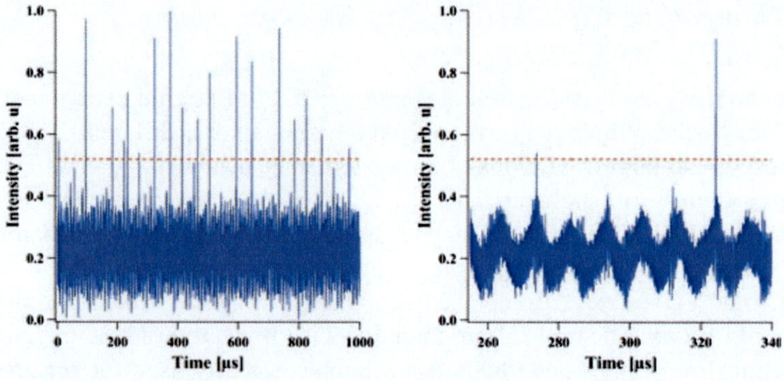

(a) (b)

Fig. 5.17 **a** Experimental time traces of the QCL under EOF and a sine-wave modulation with a period of 10 μs and an amplitude of 4 mA. The threshold corresponding to an abnormality index of 2 is displayed with a dashed line. **b** Close up on two extreme events and the low-amplitude sine modulation, revealing that extreme events can only occur on top of the periodic modulation

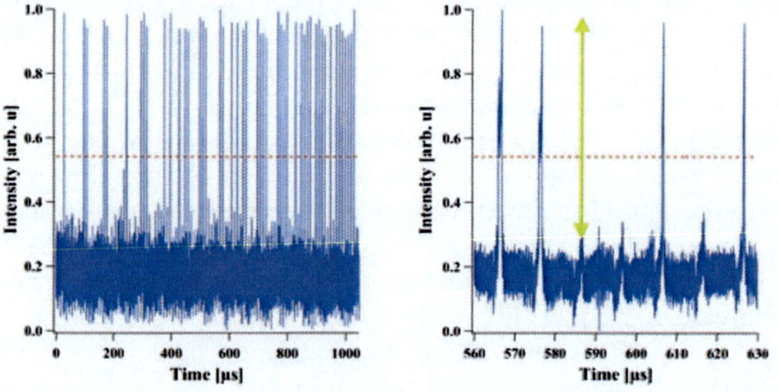

(a) (b)

Fig. 5.18 **a** Experimental time traces of the QCL under EOF and an asymmetric square-wave modulation with a period of 10 μs, an amplitude of 4 mA and a duty cycle of 20%. The threshold corresponding to an abnormality index of 2 is displayed with a dashed line. **b** Close up on four extreme events and the low-amplitude square modulation, the vertical arrow represents the amplitude discrepancy between the extreme pulses and the square-wave modulation. Extreme events are synchronized with the pulse up of the perturbation

pulse up corresponding to the upper part of the square wave. Figure 5.18b shows the synchronization of the extreme events with the periodic modulation signal and also depicts the difference in amplitude between the extreme events and one of the pulse-up which did not give birth to a magnified event. The latter case is however less frequent than in the configuration with a low-amplitude sine forcing since the success rate for extreme events is now more than 40%. It is relevant to note than this success rate is very close to that predicted in the simulations for another type of semiconductor laser [31]. Further comparison could be drawn with the laser diode

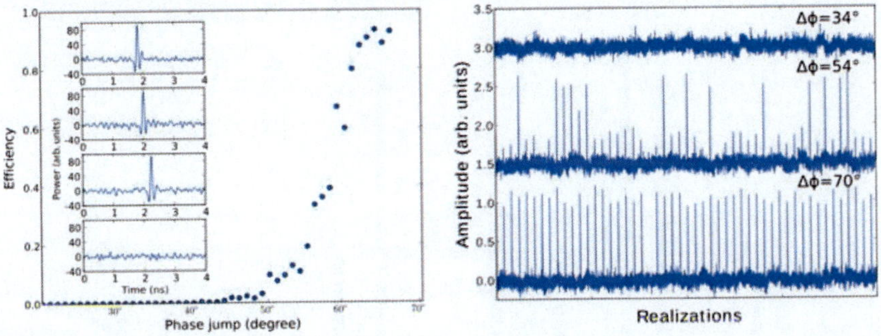

Fig. 5.19 Perturbation efficiency depending on the amplitude of the phase jump, in the case of a VCSEL subject to injection. The left panel displays the success rate as well as the associated extreme pulses. Low-amplitude phase jumps do not trigger any visible response (bottom inset, 52°), while above a certain threshold (55°) nearly all perturbations are successful. The right panel shows the response to phase jumps of the injection beam depending on the phase jump height. In all cases the phase jump last 100 ps and the repetition period is 20 ns. On the top trace, no response is observed. On the bottom trace, nearly all perturbations successfully trigger a response. The middle trace is the intermediate case in which either types of response can be observed. Courtesy of [27]

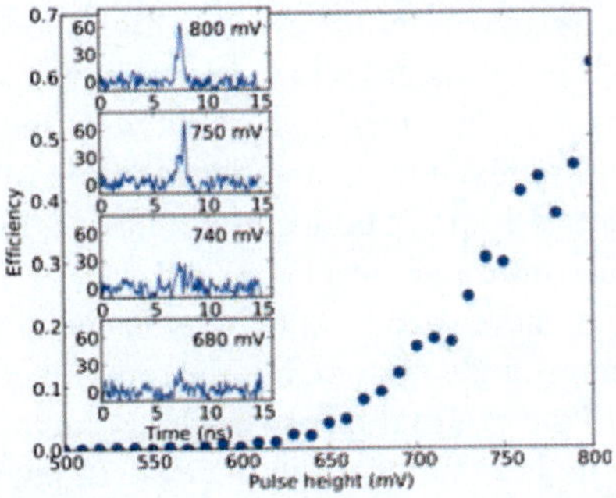

Fig. 5.20 Configuration similar to that exposed in Fig. 5.19 (left chart) but for a pulse-up operation instead of a phase-jump injection scheme. The success rate is expressed as a function of the amplitude of the pulse-up. A comparison with Fig. 5.19 (left chart) shows that the triggered extreme events have a different shape and last longer. Courtesy of [27]

case because QCLs have not been found to exhibit relaxation frequency and this might be better to favor rogue wave operations. Indeed, modulating the laser at a frequency below the relaxation frequency leads to a configuration where rogue waves are likely to occur. Conversely, modulating above the relaxation frequency does not ensure a high success rate for rogue waves. It is even possible to totally avoid rogue waves by modulating at a frequency close to the relaxation frequency [17].

The comparison can be extended to the case of VCSELs. For instance, the maximum success rate of 47% we were able to reach is in good agreement with experiments using a pulse-up configurations, since results showed a maximum value of 60%, as visualized in Fig. 5.20. And similarly to the latter case with VCSELs, the

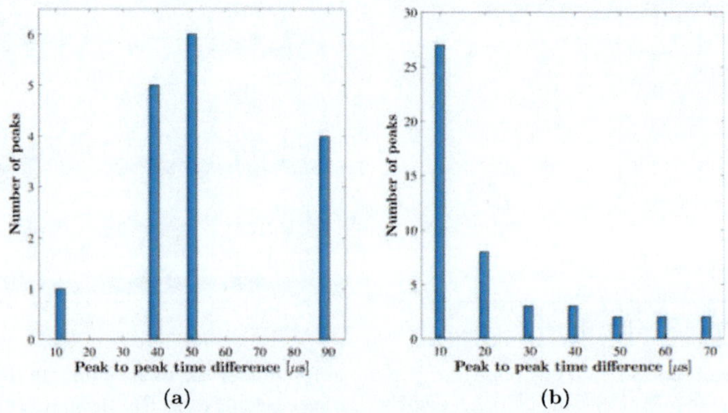

Fig. 5.21 **a** Histogram of the time-intervals between extreme events in the case where the threshold corresponding to 2 times the significant height H_s is applied to the time trace of Fig. 5.17. **b** Histogram of the time-intervals between extreme events in the case where the threshold corresponding to 2 times the significant height H_s is applied to the time trace of Fig. 5.18

extreme events that we were able to produce with a high success rate have an almost constant amplitude, which is consistent with a potential use for applications.

Figure 5.21 underlines that the time-interval between the extreme spikes in the case of Figs. 5.17 and 5.18 is given by an integer multiple of the time period of the external modulation, which is 10 μs in both configurations. In the case of Fig. 5.21a, some integer multiples are more likely to be selected (especially 40 μs, 50 μs and 90 μs) whereas some others never occur. The extreme events phenomenon seems to be dependant on a parameter of the system, but the long time-scale (a few dozens of μs) does not correspond to the external-cavity period (which is only a few ns) that is usually responsible for a repetition pattern in the case of phase-conjugate feedback [35]. In our case, this process needs further investigations to be clearly explained and, more specifically, why some time-intervals are favoured. A parallel for the very long timescales (more than 1000 times the ones found with other semiconductor lasers) could be drawn with a recent investigation on QD-lasers, which are also highly-damped semiconductor lasers [56]. These efforts proved, both experimentally and numerically, that QD-lasers can output square waves with a period of 1 μs when they are subject to injection with a specific wavelength detuning [57]. In the case of Fig. 5.21b, most of the extreme events are triggered every 10 μs, corresponding to the period of the external modulation. This eases the potential use of such on-demand bursts for sensing applications or neuron-like systems.

To further analyze the extreme events we triggered, we decided to superimpose the time traces at the peak of the extreme pulse, like we already did for rogue waves and dragon-king events. The results show that there is a huge difference between the sine perturbation and the pulse-up perturbation, as can be seen in Fig. 5.22. Indeed, the sine perturbation leads to an extreme event preceded by a dip and followed by oscillations with a lower amplitude, similarly to what is depicted in Fig. 5.15a. In the case of the pulse-up perturbation, there is dip corresponding to the beginning of

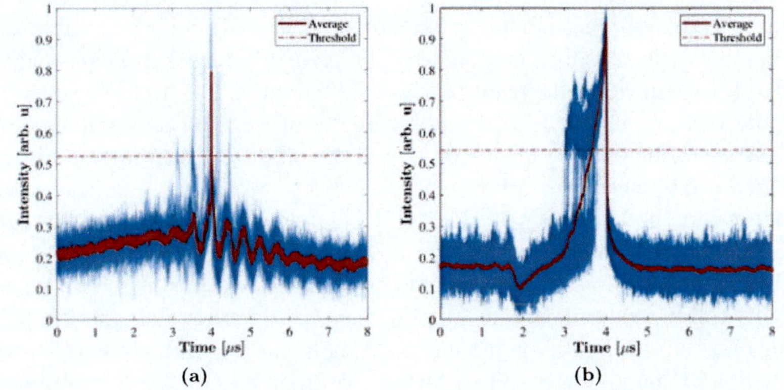

Fig. 5.22 **a** Time series centered on one single extreme event and superposition of 17 rogue waves in the case of the sine perturbation. The corresponding averaged output power is displayed with a purple solid line and the threshold corresponding to an abnormality index of 2 is displayed with a dash-dot line. **b** Same data plot but for 48 rogue waves in the case of the pulse-up perturbation

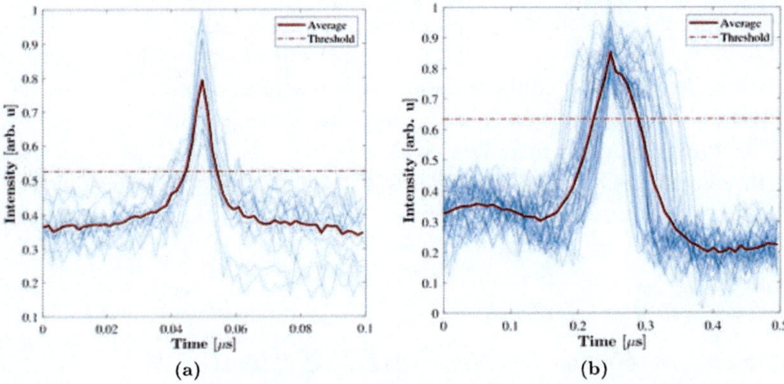

Fig. 5.23 **a** Close-up of the superposition of extreme events seen in Fig. 5.22a. **b** Close-up of the superposition of rogue waves seen in Fig. 5.15a. The two figures show that the length of the giant pulse depends on the conditions of operation

the square modulation but, after the extreme event, no oscillations can be spotted. Another difference between the several configurations studied is the width and shape of the extreme pulse. In the case of the pulse-up perturbation, Fig. 5.22b shows the extreme pulse lasts in average 500 ns and is composed of a steep incline followed by a very steep dropoff, simultaneously with the end of the 2 μs square pattern. A closer look at the pattern obtained in the case of the sine perturbation and in the case of rogue waves shows that the oscillations following the giant pulse have roughly the same frequency but the width of the pulse itself varies a lot. As can be seen in Fig. 5.23a, the spike lasts in average 20 ns in the case of the sine perturbation and 100 ns in the case of CW bias. Furthermore, the pulse is sharper in the case of a sine perturbation because the extreme events have all the same width, which is not the case with CW bias and pulse-up perturbation where the reproducibility seems hard to achieve, as depicted in Figs. 5.23b and 5.22b. It is relevant to note that, in the context of VCSELs, various patterns of extreme pulses have already been observed while varying the conditions of operation, as illustrated in Fig. 5.20. Indeed, in the case where the injection phase-jump is replaced by a pulse-up similar to the one we applied to the QCL, the VCSEL can generate extreme events lasting approximately 1 ns, which is almost ten time longer than those triggered with the phase-jump configuration. In addition, the pulse-up also seems to suppress the dip that usually follows extreme events. Turconi and co-workers explain that, on the one hand, the pulse-up configuration induces a strong linear response, or, in other words a large bump, before giving birth to a narrow extreme pulse only if the amplitude of the bump has reached a threshold. On the other hand, the phase-jump perturbation has a slight linear response and only a narrow extreme pulse is displayed, provided that the phase-jump is above the aforementioned threshold ($\Delta\phi = 55°$). Although our time traces with a pulse-up operation (Fig. 5.22b) show an initial bump followed by an extreme event, the building process seems different in the case of QCLs because the pulse-up signal is small compared to the average size of the extreme pulse, as shown by the vertical arrow in Fig. 5.18. Another conclusion of their experimental efforts is that the injection configuration allows reaching a maximum success rate (nearly 100%) higher than the pulse-up configuration (roughly 60%). The latter being compatible with the success rate (47%) we were able to achieve in the case of QCLs. Further investigation is required in order to fully understand how the shape of the extreme events varies with all the parameters we studied in the case of QCLs under EOF.

5.2.7 Extreme Events for Neuron-Like Systems

Many trials have been carried out in several fields in order to reproduce the behavior of neurons. Some of the first attempts were performed with chemical oscillators [58] but the typical characteristic time of non-linear chemical dynamics is, at least, in the order of second [59] and is thus slower than a biological neuron cluster. The electrical domain then drove attention [60] with complex grids and operation speeds below 1 ms, which is up to one hundred time faster than conventional neurons. In order to obtain a faster process, optical neurons have also been considered [61] with many configurations based on VCSEL as the core component [62] and for several experimental activation schemes such as polarisation switching [63], electrical bias stimulation [64] or via saturable absorber [65]. With VCSELs, the maximum processing rate is close to the GHz and is thus several order of magnitude faster than the best performances with electrical circuits [66]. The purpose of neuromorphic computing systems is to process information tasks inspired by the brain's powerful computational abilities with ultra-fast lasers. One of the key requirements to replicate biological neurons' response is to perform controllable inhibition and excitation (or activation) [62]. The latter is fragmented in several sub-categories called tonic spiking, phasic spiking, tonic bursting, phasic bursting, excitability, spike thresholding and spike latency among others [67]. While most of these sub-categories have already been successfully assessed in the case of VCSELs [66], we demonstrate that tonic spiking, phasic spiking and spike thresholding can be achieved in a QCL under OF, as illustrated in the following.

In the case of QCLs, we were able to trigger spikes in two different configurations. The one with the sine forcing allowed us displaying spikes with a low success rate and with uneven amplitude. Consequently, it would be difficult to take advantage of this process in order to create an optical neuron since the two key parameters for such configuration are a high reproducibility and a constant amplitude that does not depend on the excitation strength, provided that this excitation reaches a threshold. The second configuration we tested, with the pulse-up excitation, showed an improved success rate and a consistent amplitude when extreme events are triggered. The conclusion is that the latter configuration is more appropriate for neuromorphic photonics. In order to go a step beyond in the analysis of our two configurations leading to controllable extreme events, we draw a comparison with existing analysis in other semiconductor lasers. The existence of optical bursts with various heights can be related to the aforementioned optical rogue waves that we used to describe the threshold criteria. A closer look at the extreme spikes we triggered (Fig. 5.22b) shows that these bursts start with a very steep edge, then are followed by a large incline up to the maximum amplitude which is itself followed by an abrupt edge down to the free-running value. For all the extreme bursts, the width of the described process can range from 0.3 μs to 1 μs but is not identical for all the spikes. These three steps are very similar to what has been observed in the case of optical systems under either injection or optical feedback when applying the Van der Pol-Fitzhugh-Nagumo (VdPFN) model [68], consisting in a general scenario of relaxation oscillations which are characterized by the competition of two timescales: the timescale of carriers in the nanosecond range and a thermal timescale in the microsecond range for a configuration with quantum dot lasers which are highly-damped semiconductor lasers [57, 69]. In particular, Ref. [69] also shows spikes with regular amplitude and similar sharp edges, as shown in Fig. 5.24. Furthermore, a jitter is also present in the superposition diagram, even though in the case of quantum dots, the maximum of the burst occurs at the beginning of the pattern while it occurs at the end of the pattern in our case. In addition, the characteristic time of the pattern observed with quantum dots is very similar to the one we observed with the QCL under OF. The VdPFN model could thus explain why the typical time-delay dynamics, in the range of nanosecond, observed in other semiconductor laser configurations is not reproduced in our case. Last but not least, the VdPFN model has been recently studied in order to draw a qualitative comparison between semiconductor lasers with external optical feedback and biological neurons simulated with a stochastic model [70].

For each occurrence, the spike has a characteristic time of the order of the μs and that is of paramount importance in neuromorphic photonics. This indeed means that our basic optical neuron system operates 10000 times faster than biological neurons and 100 times faster than electronic artificial neurons. Compared with existing

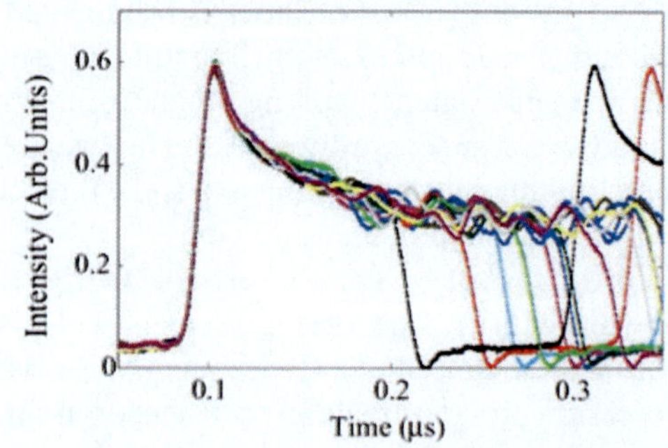

Fig. 5.24 Superposition of typical optothermal pulsations in the case of a quantum dot laser subject to feedback. The pulses exhibit both dynamics in the order of microseconds and time jittering. Courtesy of [69]

optical neuromorphic systems which use semiconductor lasers under the injection of a remote laser, our configuration takes advantage of external optical feedback and consequently, only one laser component is required to build one basic spiking neuron. This means that excitable QCLs under external optical feedback have a response time that bridges the ultra-fast neuromorphic optics obtained with VCSELs (characteristic time in the order of the nanosecond) and the electrical neural grids with a characteristic time around the millisecond. Up to date, electrical networks are more complex with configurations made of thousands of neurons, and further inquiries in the case of QCLs and photonics will be dedicated to the combination of several optical neurons to build clusters [71]. The advantage of our optical feedback configuration is that only one laser is required to process a neuron whereas the injection configuration, widely used in VCSELs, requires two lasers. Another drawback of injection operation is that during the perturbation process, the wavelength of the remote laser is modified and consequently, the injection-locking is momentary lost [64]. This means that the wavelength detuning between the master VCSEL and the slave VCSEL must be tuned adequately in order to have proper operation and this adds constraints to the conditions of operation. Finally, mid-infrared wavelengths around 6 μm are in the amide I band and the amide II band of proteins [72], which means that these wavelengths are strongly absorbed by biological tissues and are adequate for precise targeting with low collateral damage, especially in neural tissues. Consequently, surgery applications such as coagulation and ablation have already been demonstrated with QCLs emitting at such wavelength [73]. This paves the way for direct interactions between biological matter and optical neurons based on mid-infrared QCLs within the amide I and II bands.

5.3 Conclusion and Future Work

This chapter was dedicating to the experimental analysis of rogue waves and extreme events in the case of a QCL under EOF. Even though it is possible to exhibit rogue waves in the case of a free-running QCL with EOF, they are not easily predicted and extreme events are more likely to happen in a configuration with a periodic modulation of the pumping current and a small amplitude. A fine tuning of the modulation pattern and frequency allows controlling the time-interval between the spikes. Extreme events could be used in applications such as sensing, especially when they can be triggered on demand, provided that the amplitude of these events is regular and that the extra power can be precisely assessed. Indeed, the MCT detector used in this study does not measure the continuous power of the QCL and so this value cannot be compared with that of the extreme events. Having a global picture about the triggering of extreme events and the entrainment phenomenon is also essential when it comes to realize a secure transmission using two quantum cascade lasers. The transmission is actually secure because the message to be transmitted is hidden in the chaotic pattern and this message is to be recovered thanks to a chaos synchronization. However, if the chaotic pattern is disturbed by, for instance, extreme events, the synchronization can be jammed or, even worse, the message can become obvious amid the chaotic carrier. The privacy of the message would consequently be broken. This can be considered a serious threat because encoding a message introduce specific frequencies in the electrical signal driving the QCL and may lead to extreme events synchronized with the pulse-up. A thorough analysis of the amplitude and the frequency of the message is necessary to minimize such drawbacks.

Future efforts will focus on giving a precise cartography of the extreme events with respect to the amplitude of the perturbation, the period of the perturbation and the external-cavity length. Indeed, it was found in the case of a laser diode subject to phase-conjugate feedback that the length of the cavity had a preponderant role and that rogue waves could only be observed if the external cavity was short enough [74]. As many experimental and numerical works focused on the injection scheme to trigger rogue waves, it would also be relevant to study more carefully this configuration in the case of QCLs and to compare our results with the numerous modelling results if we are able to produce rogue waves. Last but not least, extreme events can be envisioned to disrupt transmissions in the context of all-optical processing of RF signals. Indeed, it would be interesting to analyze the impact of such sudden bursts in radio-over-fiber links [75] or to inject these bursts in free-space optics (FSO) systems in order to degrade the BER of the channel up to values where the message cannot be surely recovered.

References

1. Kharif C, Pelinovsky E (2003) Physical mechanisms of the rogue wave phenomenon. Eur J Mechan-B/Fluids 22(6):603–634
2. Perkins S (2006) Dashing rogues: Freak ocean waves pose threat to ships, deep-sea oil platforms. Sci News 170(21):328–329
3. Didenkulova I, Slunyaev A, Pelinovsky E, Kharif C (2006) Freak waves in 2005. Nat Hazards Earth Syst Sci 6(6):1007–1015
4. Longuet-Higgins MS (1952) On the statistical distributions of the heights of sea waves. J Maritime Res 9:245–266

5. Donelan MA, Drennan WM, Magnusson AK (1996) Nonstationary analysis of the directional properties of propagating waves. J Phys Oceanogr 26(9):1901–1914
6. Mori N, Liu PC, Yasuda T (2002) Analysis of freak wave measurements in the Sea of Japan. Ocean Eng 29(11):1399–1414
7. Chabchoub A, Hoffmann N, Onorato M, Akhmediev N (2012) Super rogue waves: observation of a higher-order breather in water waves. Phys Rev X 2(1):011015
8. Donelan MA, Magnusson A-K (2017) The making of the Andrea wave and other rogues. Sci Rep 7:44124
9. Yan Z (2011) Financial rogue waves appearing in the coupled nonlinear volatility and option pricing model, *arXiv preprint* arXiv:1101.3107
10. Ganshin A, Efimov V, Kolmakov G, Mezhov-Deglin L, McClintock PV (2008) Observation of an inverse energy cascade in developed acoustic turbulence in superfluid helium. Phys Rev Lett 101(6):065303
11. El-Labany S, Moslem W, El-Bedwehy N, Sabry R, El-Razek HA (2012) Rogue wave in Titan's atmosphere. Astrophys Space Sci 338(1):3–8
12. Frolov NS, Grubov VV, Maksimenko VA, Lüttjohann A, Makarov VV, Pavlov AN, Sitnikova E, Pisarchik AN, Kurths J, Hramov AE (2019) Statistical properties and predictability of extreme epileptic events. Sci Rep 9(1):7243
13. Onorato M, Residori S, Bortolozzo U, Montina A, Arecchi F (2013) Rogue waves and their generating mechanisms in different physical contexts. Phys Rep 528(2):47–89
14. Dean R (1990) Freak waves: a possible explanation. In: Water wave kinematics. Springer, pp 609–612
15. Janssen PA (2003) Nonlinear four-wave interactions and freak waves. J Phys Oceanogr 33(4):863–884
16. Bonatto C, Feyereisen M, Barland S, Giudici M, Masoller C, Leite JRR, Tredicce JR (2011) Deterministic optical rogue waves. Phys Rev Lett 107(5):053901
17. Perrone S, Vilaseca R, Zamora-Munt J, Masoller C (2014) Controlling the likelihood of rogue waves in an optically injected semiconductor laser via direct current modulation. Phys Rev A 89(3):033804
18. Kovalsky MG, Hnilo AA, Tredicce JR (2011) Extreme events in the Ti: sapphire laser. Opt Lett 36(22):4449–4451
19. Solli D, Ropers C, Koonath P, Jalali B (2007) Optical rogue waves. Nature 450(7172):1054
20. Montina A, Bortolozzo U, Residori S, Arecchi F (2009) Non-Gaussian statistics and extreme waves in a nonlinear optical cavity. Phys Rev Lett 103(17):173901
21. Hammani K, Finot C, Dudley JM, Millot G (2008) Optical rogue-wave-like extreme value fluctuations in fiber Raman amplifiers. Optics Exp 16(21):16 467–16 474
22. Churkin DV, Gorbunov OA, Smirnov SV (2011) Extreme value statistics in Raman fiber lasers. Opt Lett 36(18):3617–3619
23. Pisarchik AN, Jaimes-Reátegui R, Sevilla-Escoboza R, Huerta-Cuellar G, Taki M (2011) Rogue waves in a multistable system. Phys Rev Lett 107(27):274101
24. Lecaplain C, Grelu P, Soto-Crespo J, Akhmediev N (2012) Dissipative rogue waves generated by chaotic pulse bunching in a mode-locked laser. Phys Rev Lett 108(23):233901
25. Schires K, Hurtado A, Henning I, Adams M (2012) Rare disruptive events in polarisation-resolved dynamics of optically injected 1550 nm VCSELs. Electron Lett 48(14):872–874
26. Zamora-Munt J, Garbin B, Barland S, Giudici M, Leite JRR, Masoller C, Tredicce JR (2013) Rogue waves in optically injected lasers: origin, predictability, and suppression. Phys Rev A 87(3):035802
27. Turconi M, Garbin B, Feyereisen M, Giudici M, Barland S (2013) Control of excitable pulses in an injection-locked semiconductor laser. Phys Rev E 88(2):022923
28. Ahuja J, Nalawade DB, Zamora-Munt J, Vilaseca R, Masoller C (2014) Rogue waves in injected semiconductor lasers with current modulation: role of the modulation phase. Optics Exp 22(23):28 377–28 382
29. Hurtado A, Javaloyes J (2015) Controllable spiking patterns in long-wavelength vertical cavity surface emitting lasers for neuromorphic photonics systems. Appl Phys Lett 107(24):241103

30. Alvarez NM, Borkar S, Masoller C (2017) Predictability of extreme intensity pulses in optically injected semiconductor lasers. Eur Phys J Special Topics 226(9):1971–1977
31. Jin T, Siyu C, Masoller C (2017) Generation of extreme pulses on demand in semiconductor lasers with optical injection. Opt Exp 25(25):31 326–31 336
32. Reinoso JA, Zamora-Munt J, Masoller C (2013) Extreme intensity pulses in a semiconductor laser with a short external cavity. Phys Rev E 87(6):062913
33. Choi D, Wishon MJ, Barnoud J, Chang C, Bouazizi Y, Locquet A, Citrin D (2016) Low-frequency fluctuations in an external-cavity laser leading to extreme events. Phys Rev E 93(4):042216
34. Uy C-H, Rontani D, Sciamanna M (2017) Vectorial extreme events in VCSEL polarization dynamics. Opt Lett 42(11):2177–2180
35. Karsaklian Dal Bosco A, Wolfersberger D, Sciamanna M (2013) Extreme events in time-delayed nonlinear optics. Opt Lett 38(5):703–705
36. Mercier É, Even A, Mirisola E, Wolfersberger D, Sciamanna M (2015) Numerical study of extreme events in a laser diode with phase-conjugate optical feedback. Phys Rev E 91(4):042914
37. Lee MW, Baladi F, Burie JR, Bettiati MA, Boudrioua A, Fischer AP (2015) Observation of rogue waves in a 980nm-laser diode subject to filtered optical feedback. In: Conference on lasers and electro-optics/pacific rim. Optical Society of America, p 25B2_4
38. Lin SS, Hwang SK, Liu JM (2015) High-power noise-like pulse generation using a 1.56-μm all-fiber laser system. Opt Exp 23(14):18 256–18 268
39. Soriano MC, Ortín S, Brunner D, Larger L, Mirasso CR, Fischer I, Pesquera L (2013) Opto-electronic reservoir computing: tackling noise-induced performance degradation. Opt Exp 21(1):12–20
40. Zamora-Munt J, Mirasso CR, Toral R (2014) Suppression of deterministic and stochastic extreme desynchronization events using anticipated synchronization. Phys Rev E 89(1):012921
41. Seo D, Park J, McInerney J, Osinski M (1988) Effects of feedback asymmetry in external-cavity semiconductor laser systems. Electron Lett 24(12):726–728
42. Residori S, Bortolozzo U, Montina A, Lenzini F, Arecchi F (2012) Rogue waves in spatially extended optical systems. Fluct Noise Lett 11(01):1240014
43. Sornette D (2009) Dragon-kings, black swans and the prediction of crises. *arXiv preprint* arXiv:0907.4290
44. Gil L, Sornette D (1996) Landau-Ginzburg theory of self-organized criticality. Phys Rev Lett 76(21):3991
45. Ancey C (2012) Are there "dragon-kings" events (ie genuine outliers) among extreme avalanches? Eur Phys J Special Topics 205(1):117–129
46. Cavalcante HL, Oriá M, Sornette D, Ott E, Gauthier DJ (2013) Predictability and suppression of extreme events in a chaotic system. Phys Rev Lett 111(19):198701
47. Mishra A, Saha S, Vigneshwaran M, Pal P, Kapitaniak T, Dana SK (2018) Dragon-king-like extreme events in coupled bursting neurons. Phys Rev E 97(6):062311
48. Sornette D, Ouillon G (2012) Dragon-kings: mechanisms, statistical methods and empirical evidence. Eur Phys J Special Topics 205(1):1–26
49. Martinerie J, Adam C, Le Van Quyen M, Baulac M, Clemenceau S, Renault B, Varela FJ (1998) Epileptic seizures can be anticipated by non-linear analysis. Nat Med 4(10):1173
50. Feigenbaum JA (2001) A statistical analysis of log-periodic precursors to financial crashes. Quant Finan 1:346–360
51. Laptev N, Yosinski J, Li LE, Smyl S (2017) Time-series extreme event forecasting with neural networks at uber. In: International conference on machine learning, vol 34, pp 1–5
52. Dakos V, Scheffer M, van Nes EH, Brovkin V, Petoukhov V, Held H (2008) Slowing down as an early warning signal for abrupt climate change. In: Proceedings of the national academy of sciences, vol 105, no 38, pp 14308–14312
53. Robertson J, Deng T, Javaloyes J, Hurtado A (2017) Controlled inhibition of spiking dynamics in VCSELs for neuromorphic photonics: theory and experiments. Opt Lett 42(8):1560–1563
54. Granese NM, Lacapmesure A, Agüero MB, Kovalsky MG, Hnilo AA, Tredicce JR (2016) Extreme events and crises observed in an all-solid-state laser with modulation of losses. Opt Lett 41(13):3010–3012

55. Marino F, Giudici M, Barland S, Balle S (2002) Experimental evidence of stochastic resonance in an excitable optical system. Phys Rev Lett 88(4):040601
56. Huang H (2017) Optical nonlinearities in quantum dot lasers for high-speed communications, PhD dissertation, Télécom ParisTech
57. Dillane M, Tykalewicz B, Goulding D, Garbin B, Barland S, Kelleher B (2019) Square wave excitability in quantum dot lasers under optical injection. Opt Lett 44(2):347–350
58. Yang L, Dolnik M, Zhabotinsky AM, Epstein IR (2000) Oscillatory clusters in a model of the photosensitive Belousov-Zhabotinsky reaction system with global feedback. Phys Rev E 62(5):6414
59. Epstein IR, Showalter K (1996) Nonlinear chemical dynamics: oscillations, patterns, and chaos. J Phys Chem 100(31):13132–13147
60. Benjamin BV, Gao P, McQuinn E, Choudhary S, Chandrasekaran AR, Bussat J-M, Alvarez-Icaza R, Arthur JV, Merolla PA, Boahen K (2014) Neurogrid: a mixed-analog-digital multichip system for large-scale neural simulations. Proc IEEE 102(5):699–716
61. Mos EC, Hoppenbrouwers JJ, Hill MT, Blum MW, Schleipen JJ et al (2000) Optical neuron by use of a laser diode with injection seeding and external optical feedback. IEEE Trans Neural Netw 11(4):988–996
62. Nahmias MA, Shastri BJ, Tait AN, Prucnal PR (2013) A leaky integrate-and-fire laser neuron for ultrafast cognitive computing. IEEE J Sel Top Quantum Electron 19(5):1–12
63. Hurtado A, Schires K, Henning I, Adams M (2012) Investigation of vertical cavity surface emitting laser dynamics for neuromorphic photonic systems. Appl Phys Lett 100(10):103703
64. Robertson J, Wade E, Hurtado A (2019) Electrically controlled neuron-like spiking regimes in vertical-cavity surface-emitting lasers at ultrafast rates. IEEE J Sel Top Quantum Electron 25(6):1–7
65. Zhang Y, Xiang S, Gong J, Guo X, Wen A, Hao Y (2018) Spike encoding and storage properties in mutually coupled vertical-cavity surface-emitting lasers subject to optical pulse injection. Appl Opt 57(7):1731–1737
66. Robertson J, Wade E, Kopp Y, Bueno J, Hurtado A (2019) Toward neuromorphic photonic networks of ultrafast spiking laser neurons. IEEE J Sel Top Quantum Electron 26(1):1–15
67. Izhikevich EM (2004) Which model to use for cortical spiking neurons? IEEE Trans Neural Netw 15(5):1063–1070
68. Marino F, Catalán G, Sánchez P, Balle S, Piro O (2004) Thermo-optical "canard orbits" and excitable limit cycles. Phys Rev Lett 92(7):073901
69. Tierno A, Radwell N, Ackemann T (2011) Low-frequency self-pulsing in single-section quantum-dot laser diodes and its relation to optothermal pulsations. Phys Rev A 84(4):043828
70. Tiana-Alsina J, Quintero-Quiroz C, Masoller C (2019) Comparing the dynamics of periodically forced lasers and neurons. New J Phys 21(10):103039
71. Dolcemascolo A, Miazek A, Veltz R, Marino F, Barland S (2020) "Effective low-dimensional dynamics of a mean-field coupled network of slow-fast spiking lasers." Phys Rev E 101(5):052208
72. Edwards G, Logan R, Copeland M, Reinisch L, Davidson J, Johnson B, Maciunas R, Mendenhall M, Ossoff R, Tribble J et al (1994) Tissue ablation by a free-electron laser tuned to the amide II band. Nature 371(6496):416–419
73. Huang Y, Kang JU (2012) Corneal tissue ablation using 6.1 μm quantum cascade laser. In: Ophthalmic Technologies XXII, vol 8209. International Society for Optics and Photonics, p 82091W
74. Karsaklian Dal Bosco A (2013) Chaos and high-frequency self-pulsations in a laser diode with phase-conjugate feedback, PhD dissertation, CentraleSupélec
75. Hung Y-H, Hwang S-K (2013) Photonic microwave amplification for radio-over-fiber links using period-one nonlinear dynamics of semiconductor lasers. Opt Lett 38(17):3355–3358

Chapter 6
Conclusions and Perspectives

In this dissertation, we have focused on QCLs emitting in the mid-infrared domain since this range of wavelengths is of prime importance for free-space communications due to the high transparency of the atmosphere. The mid-infrared domain is also more adequate than near-infrared wavelengths because detrimental factors, such as divergence and scintillation, are reduced at higher wavelengths. Other applications involving QCLs include, but are not limited to, countermeasure systems, surgery and high-precision spectroscopy because many molecules and chemical compounds have a strong absorption in the mid-infrared. Our efforts were directed towards the implementation of a free-space secure communication with chaotic QCLs, thus requiring a thorough analysis of non-linear phenomena in this kind of semiconductor lasers.

Chaos experiments with a QCL under EOF determined the best conditions of operation in order to have complex and sustained chaos at mid-infrared wavelength. We showed that chaos could be generated not only close to the injection threshold, but also far from threshold where the QCL is able to output more optical power. We also showed that the lower the temperature, the easier the generation of chaotic patterns, but cryogenic temperatures come with bulky setups and may not be suitable for real-field experiments requiring versatility. We thus decided to focus on experiments with Peltier-cooled QCLs operated far above threshold to maximize output power and compactness. We also showed that chaos was more developed in the case of a quasi-continuous bias but once again, this may not be suitable in the case of secure communications. This experimental work was complemented by a numerical analysis of the influence of the α-factor on the bifurcation diagram of the QCL under EOF. We showed that despite the fact that the reciprocal space curvature of the subbands in a QCL should give a LEF very close to zero, only a quite large α-factor can be compatible with the dynamics we observed. These findings were backed by a self-mixing interferometry derivation of the experimental α-factor of a QCL pumped far above threshold, which confirms a larger value than expected. Further studies will be required to determine if the value of the LEF depends on the method used to

O. Spitz, *Mid-infrared Quantum Cascade Lasers for Chaos Secure Communications*, Springer Theses, https://doi.org/10.1007/978-3-030-74307-9_6

derive it. Indeed, experimental methods focusing on low-frequency schemes (e.g. self-mixing interferometry) tend to give higher LEF values than those relying on multi-GHz analysis. This may imply that the LEF is linked to an opto-thermal effect and that high values of LEF are restrained to a few dozens of MHz, where we experimentally retrieved chaotic patterns. This could explain why we did not observe non-linear dynamics at GHz frequencies, even though QCLs do not exhibit relaxation oscillations and should thus be prone to very high-frequency dynamics.

To deepen our understanding of the chaotic phenomena, we decided to further tailor the bias current of the QCL under EOF by adding a periodic bias modulation, as already experimented in other semiconductor lasers. This combination of EOF and bias forcing allowed us confirming that QCLs are prone to the entrainment phenomenon. The latter drives the chaotic spiking in order to synchronize it with the periodic perturbation and can offer means of communication with controlled chaos. The entrainment phenomenon was proven effective in both the case of a sine perturbation and a square perturbation, and an accurate control of the number of spikes per period was achieved through the tuning of the frequency and of the amplitude of the periodic forcing. Further tuning of the QCL's output can be considered with feedback techniques we have not implemented yet. For instance, filtered optical feedback and phase-conjugate feedback could help us generating complex chaotic patterns at high frequency, even if the peculiarities of mid-infrared wavelengths brake the implementation of such setups.

Other non-linear phenomena were also observed with QCLs. For instance, it is possible to output a square wave with cross-polarization reinjection technique. The period and duty cycle of that pattern can be tuned by varying the feedback strength and the polarization of the reinjected beam. This is quite unusual in other semiconductor lasers where a cross-polarization technique leads to a square-wave emission but with a fixed period related to the length of the external cavity. This result can be of paramount importance in the case of all-optical modulation of QCLs albeit an accurate lever must be found to control the period and duty cycle of such pattern. In order to increase the bandwidth of chaos in QCLs, it is possible to take advantage of a combination between external optical injection and EOF. Indeed, the results obtained when injecting the light from another QCL showed that we were able to output higher non-linear frequencies than with EOF only. Further studies will try to generate chaos with frequencies of several hundreds of MHz of bandwidth, because this means that the frequency of the message to be transmitted can be higher while still well hidden among the chaotic carrier. Another option to achieve high-frequency modulation of a QCL would be to inject in the QCL's emitting facet with a near-infrared laser because this was proven useful in quenching the QCL at high rates. We will carry out experiments to determine the influence when injecting a wavelength that is strongly different from the one emitted by the QCL, in order to determine the minimum rise-time achievable with this method.

Eventually, we also discovered rogue waves and extreme events in the output of a QCL under EOF, as already observed in other semiconductor lasers. Although semiconductor lasers can be seen as a test-bed for rogue waves, which are found in many fields of physics such as hydrodynamics, acoustics or astrophysics, this phenomenon can also be hazardous because the high-power bursts can deteriorate optical components. Moreover, they can also disturb a transmission in the case where a message is enciphered within a chaotic carrier. We thus described precursors of rogue waves in order to forecast them. We also showed it was possible to trigger them on-demand with an appropriate pulse-up modulation and this means that extreme events could be used in order to jam detectors. Further efforts will study how to suppress rogues once the precursors have been identified. This may be easier in the case of QCLs under EOF because the typical time of such events is in the order of 100 ns, which is 1000 times longer that observed in laser diodes.

All these insights about the non-linear dynamics in QCLs drove us to an experimental demonstration of chaos synchronization in two QCLs sharing similar parameters. We showed that the slave QCL could output a signal either synchronized or anti-synchronized with that of the master laser, similarly to what was unveiled for laser diodes and VCSELs. We then combined the chaotic signal of the master with a message at 0.5 Mbits/s to realize a secure free-space transmission at mid-infrared wavelength. The best BER was found to be 6% in the case of anti-synchronization, and a careful analysis of the eye diagrams led us to the conclusion that a potential eavesdropper could not decipher the message when getting the signal with the message embedded in chaos. This paves the way for transmission at mid-infrared wavelength, taking advantage of the stealth of the wavelength and a simpler implementation compared to scheme involving quantum key distribution.

Further efforts will thoroughly detail the advantages and drawbacks of QCLs compared to other existing sources for free-space communications. The case of ICLs is extremely relevant because these optical sources combine the advantages of mid-infrared wavelength up to 7 μm and of interband transitions. Other lasers with such transitions have already been proven to output chaos at several Gbits/s. However, no studies about the non-linear dynamics of ICLs exist and consequently, a careful analysis of all the experimental parameters, similar to the one we have carried out for QCLs, is required before comparing the real-field beam stability of near-infrared, ICL and QCL sources.

In addition, we plan to complement our transmission studies with direct-modulation schemes without chaos encryption, for both QCLs and ICLs. Several encoding methods will be tested such as OOK modulation and PAM schemes. We want to establish the first ever FSO link with a QCL/ICL directly modulated at a bit-rate of nearly 1 Gbits/s and for a 4–5 km horizontal path. The experiments will be performed in winter with low visibility and in summer with strong turbulence. Comparisons will be drawn with other semiconductor lasers emitting both in the near-infrared and mid-infrared domain in order to select the most relevant option in terms of data transmission rates and ranges in the worst weather conditions. We already proved that the choice of such a wavelength has a strong impact on the telecom channel availability, especially in the case of mizzle or fog [1]. Real-field experiments with a large incline between the emitter and the detector are also envisioned in order to assess the potential of our methods for ground-satellite transmissions. These efforts will take support on existing facilities and prior studies carried out by ONERA with laser diodes [2].

As we are planning to increase the range of our communication experiments, we will be interested in optical sources with more optical power. In this situation, BA-QCLs can be of prime interest because we do not require Gaussian beams in our experiments. We showed that it is possible to increase the output power of such lasers with weak optical feedback. It would be relevant to explore more carefully the pioneering non-linear dynamics experiments with BA-QCLs [3], as well as external modulation schemes. Thus, we would access long-range mid-infrared sources for both secure communications and regular high-speed transmissions. High-power QCLs can also be produced with Talbot coupling in the case of laser arrays [4]. The numerical non-linear analysis we performed gave details about the regimes we would be able to produce but external modulation of such arrays has not been carried out yet. We plan to experimentally realize the first QCL external-Talbot cavity in order to obtain a more versatile scheme than the integrated Talbot cavities already demonstrated. Then, the next step will be to extend our non-linear efforts to such high-power sources and compare the results with those obtained with conventional and BA one-ridge QCLs.

When taking a step back at our experimental results, some questions remain unanswered and new experiments, as well as further numerical studies, will be required. For instance, some of the non-linear phenomena we observed (all-optical square modulation, rogue waves...) are also retrieved in laser diodes but with a huge difference in terms of characteristic time. As we used the Lang and Kobayashi model to derive our numerical analysis, this may imply that we have to redesign our model by taking into account the last insights from the quantum dot community, which also found larger time-scales in the case of quantum dots under external optical injection. This will be an opportunity to close the gap between QCLs and QD-lasers which are both highly-damped semiconductor lasers but which are rarely compared in numerical analysis.

References

1. Sauvage C, Robert C, Sorrente B, Grillot F, Erasme D (2019) Study of short and mid-wavelength infrared telecom links performance for different climatic conditions. In: Environmental effects on light propagation and adaptive systems II, vol 11153. International Society for Optics and Photonics, p 111530I
2. Montmerle-Bonnefois A, Petit C, Lim C, Sauvage J-F, Meimon S, Perrault P, Mendez F, Fleury B, Montri J, Conan J-M et al (2019) Adaptive optics precompensation of a GEO feeder link: the FEEDELIO experiment. In: Applications of lasers for sensing and free space communications. Optical Society of America, pp LTh1B–3
3. Newell T, Grillot F, Gavrielides A, Kaspi R, Lu C, Yang C, Bate T, Luong S (2018)Experimental investigation of broad area quantum cascade lasers under external feedback. Opt Express 26(14):17 927–17 935
4. Gavrielides A, Newell T (2020) Global coupling of QCLs: inclusion of dynamics. Opt Express 28(6):7746–7758